D1741406

SCIENCE
IN A TOPIC
SHIPS

by Doug Kincaid
and Peter S. Coles
Illustrated by Chris Hoggett

MONTGOMERY COMBINED SCHOOL

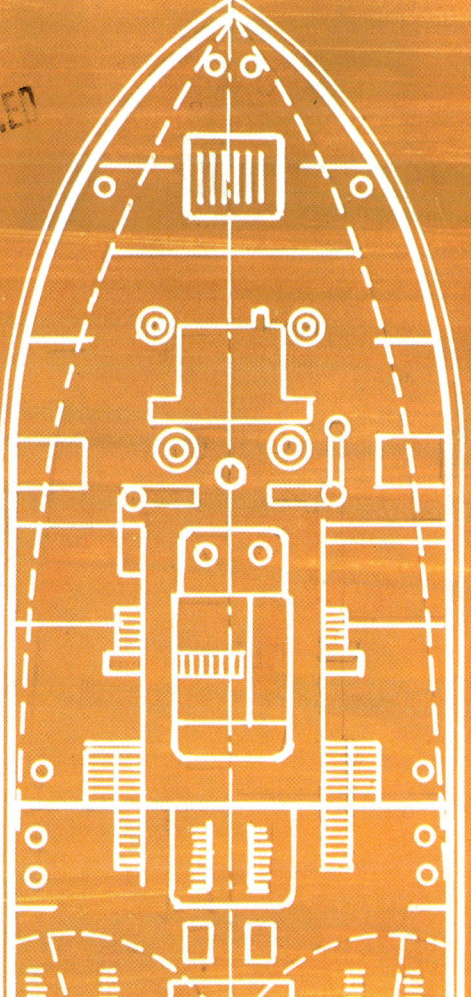

HULTON · LONDON

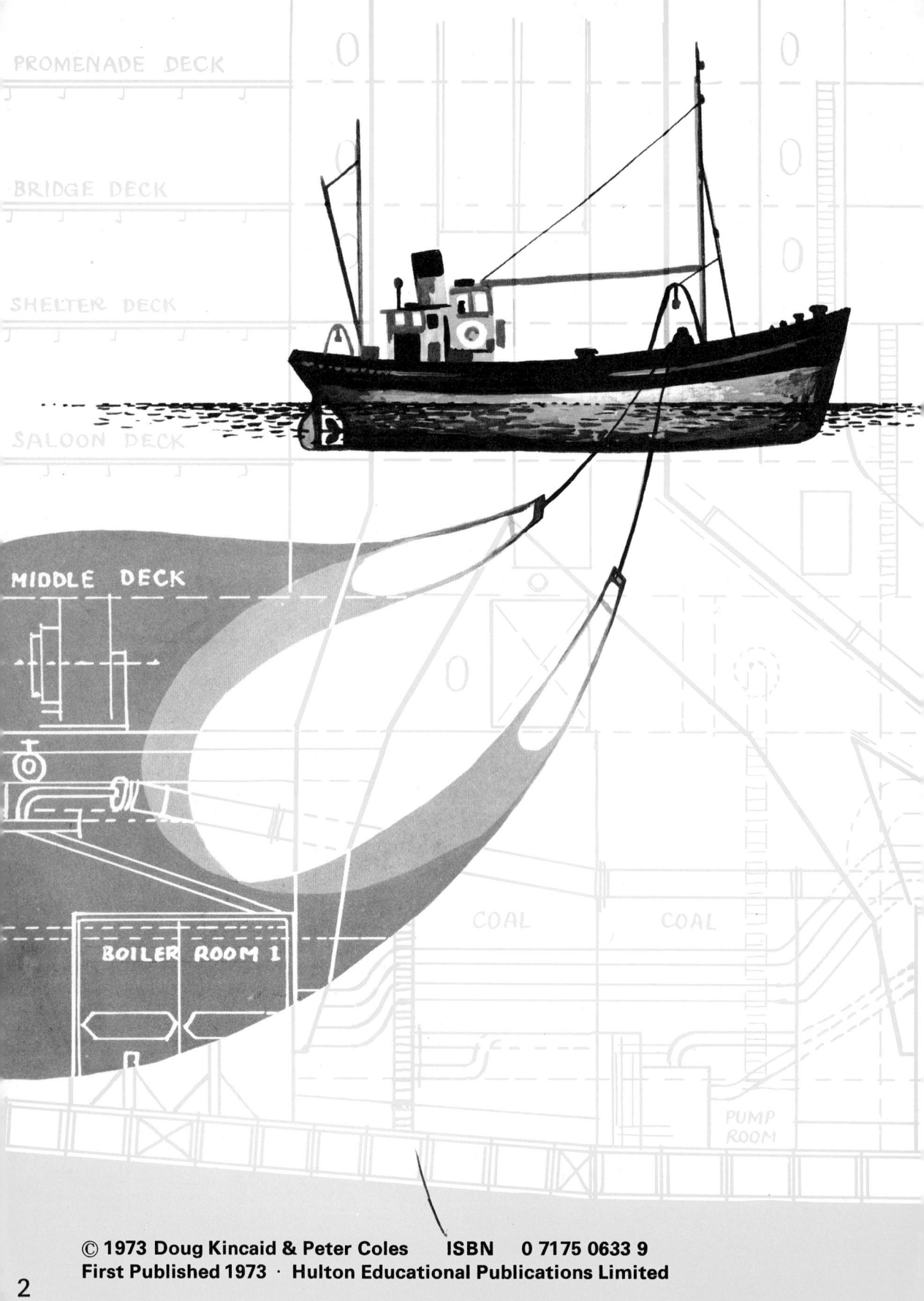

PROMENADE DECK

BRIDGE DECK

SHELTER DECK

SALOON DECK

MIDDLE DECK

BOILER ROOM 1

COAL   COAL

PUMP
ROOM

© 1973 Doug Kincaid & Peter Coles   ISBN   0 7175 0633 9
First Published 1973 · Hulton Educational Publications Limited

SCIENCE IN A TOPIC
SHIPS

About this book

This book is different from most others because :—

1. It is not complete, but only part of a study — the science part. There will be a need to use many other books to find out about other aspects of the topic — History, Geography . . .

2. It will not tell you information but will only ask you questions and suggest ways that you might find the answers for yourself.
   Many of the suggestions were some children's ways of trying to find an answer — you may have better ideas.

3. It is hoped that arising from these questions other questions will occur to you — do pursue these. (Your own questions and the ways you find to answer them are really the most important.)

4. You do not need to work through the book in the order set out; the sections of work can be done in the order that you wish.

5. There is no need to complete all of one section. If the work becomes harder as you progress through a section, see how far you can go.

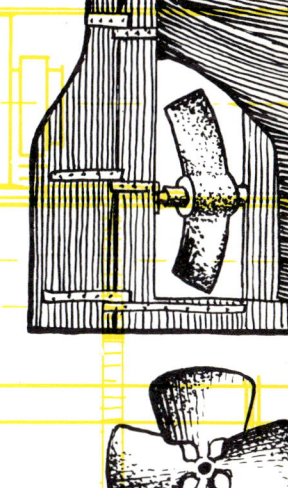

CONTENTS

Science in a Topic Series

by Doug Kincaid, Bucks County Staff Advisory Teacher, Science and
Peter S. Coles, B.Sc., Science Adviser, Buckinghamshire.

Other titles   Houses and Homes
                Bridges
                Communications
                Clothes and Costumes
                Food
                Land Transport

Published by Hulton Educational Publications Ltd., Raans Rd., Amersham, Bucks.
Photolitho and printed by Colour Reproductions Ltd., Billericay, Essex.

Why do Ships FLOAT?

SECTION ONE

Wooden Boats

The ships on this page are all made from wood. It probably seems reasonable that all wooden boats float.

An Arab coastal vessel

A Viking longship

A thirteenth-century medieval sailing ship: note the 'castles'

A large 4-mast 'man-of-war' – sixteenth century

Iron Ships

Does it seem surprising that a great mass of iron such as the QE2 floats? "An iron ship float? The man must be mad!" – such were the comments made about Brunel, the builder of one of the first iron ships – "The Great Eastern".

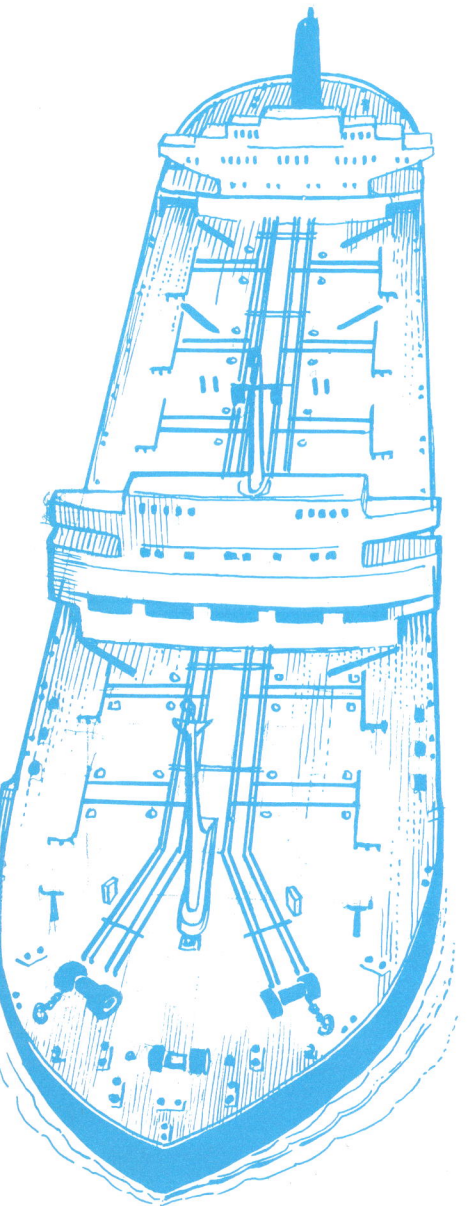

above: The Great Eastern.

left: one of the largest ships of today – the modern oil tanker.

below: H.M.S. Ark Royal
Queen Elizabeth II – both large ships made from iron.

How is it that such great iron masses float? Investigate how things float and sink.

What things float?
What things sink?

Collecting Sets

Teacher may provide certain things which should make your experiments more interesting.

Some suggestions for floating and sinking collections :–

1 Woods a) different shapes and sizes :

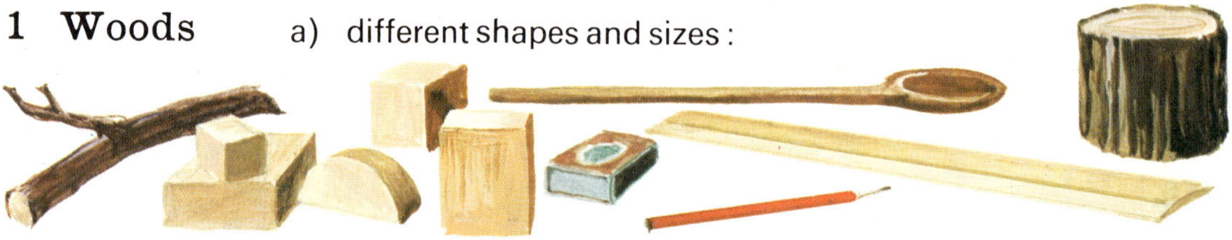

b) different kinds of wood :

Mahogany

Lime

Ash

Beech

Pitch Pine

Spruce

perhaps you could find other kinds

2 Metals a) different forms of metal :

b) different kinds of metal :

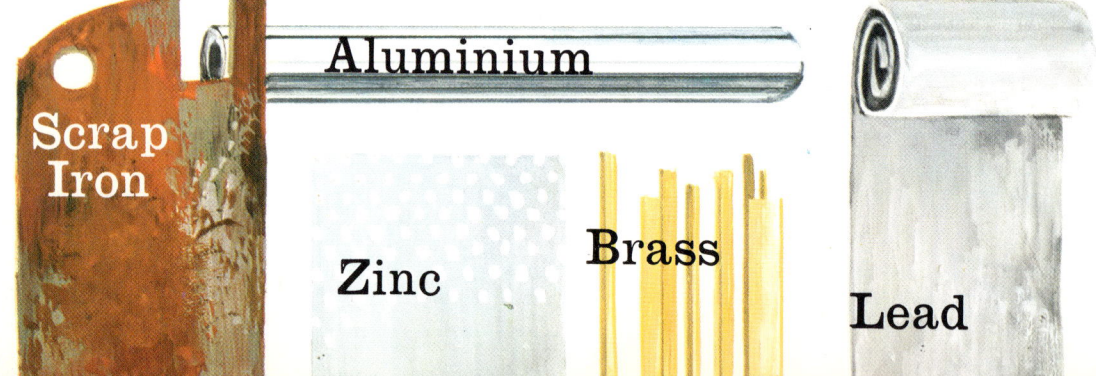

Aluminium

Scrap Iron

Zinc

Brass

Lead

3 Stones, Pebbles and Bricks

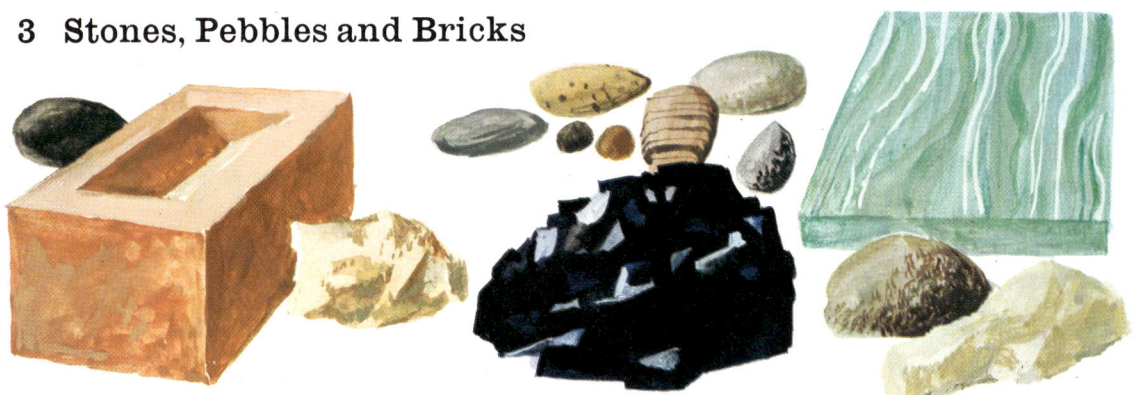

4 Glass

5 Plastics

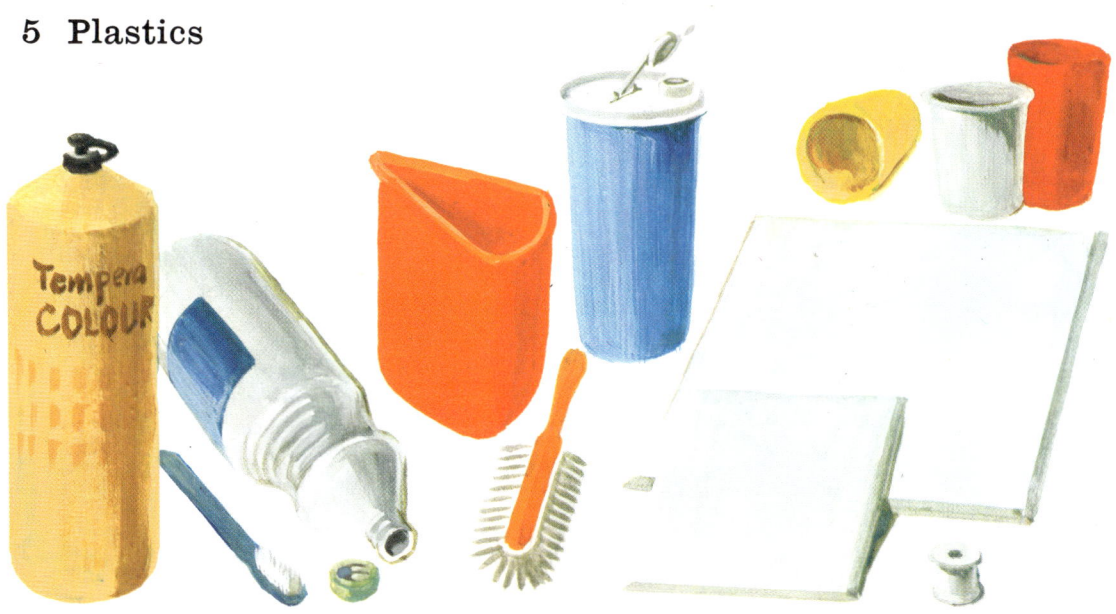

Some Research for you

You can experiment in a bucket, bowl or a transparent tank.

1 Which materials/objects float and which sink?

2 Does any material float in one form but sink in another? Look carefully at the floating set. How do they float? Always the same way up? All to the same depth? (Look particularly at the woods.)

3 Can you make any of the floating set sink?

4 Can you make any of the sinking set float?

5 Does time affect the floating set? (Find out about the Ra Expedition.)

6 Make a ball of plasticine – will it float?
Can you make it into a shape that will float?

If any float, try loading them with 'cargo'.
What mass will they carry before sinking?

washers and coins for cargo

7 Continue your investigations with aluminium foil. Try a piece 10 cm square and fold as many times as you can. Does this float or sink? Now make a boat-like shape with a similar square.

You could continue your research with different sizes and shapes of foil boats.

8 Investigate with blocks of different materials that are all the same size.

Does this float or sink? Will it carry a load? What is the greatest load it will carry?

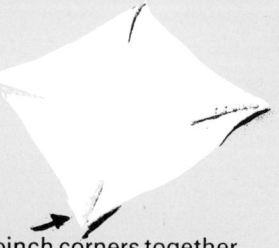

pinch corners together

Obtain a tin and a piece of iron of equal mass.

What happens when placed in water?

100g

Record your Investigations

Have you come to any conclusions as to why the floating set floats and the sinking set sinks?

1 Is it anything to do with shape?
2 Is it anything to do with size?
3 Is it anything to do with such qualities as being shiny, smooth, soft or——?

4 Is it anything to do with the material —
Does all wood float?
Does all metal sink?

What quality do you think decides whether an object floats or sinks?

Measuring

Prepare a stone ready to hang on your spring balance.

First take a balance reading in air and then repeat the experiment with your stone in water as shown.

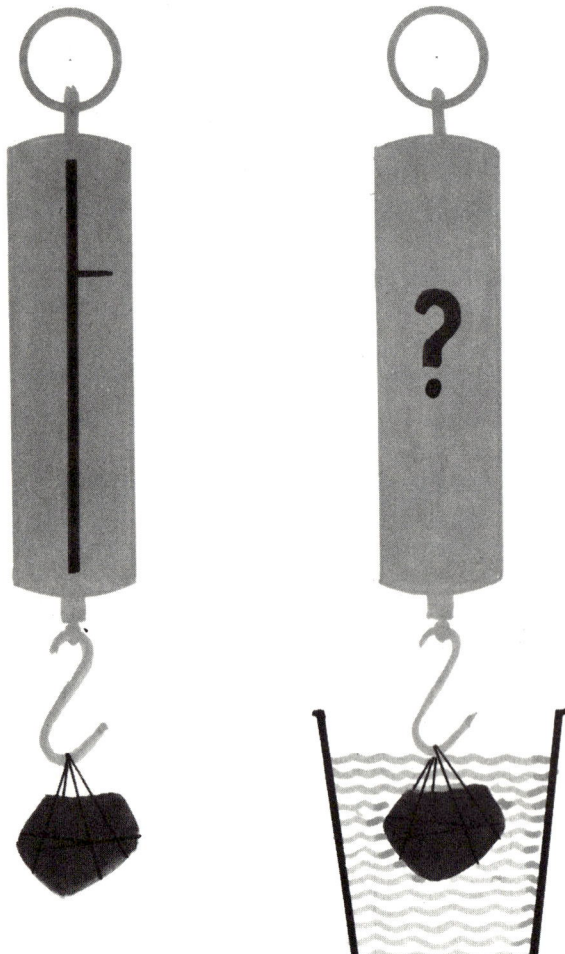

LORD KELVIN

Reproduced by kind permission of the University of Glasgow Art Collections.

Lord Kelvin, a famous scientist, once said :

"When you can measure what you are speaking about and express it in numbers, you know something about it."

To really understand how things float and sink we ought to make some measurements.

You can record your results like this :

You could continue your investigations with other objects such as :—
  bricks
  wooden blocks
  1 kg, 2 kg, 5 kg, 10 kg masses
  heavy bottles (filled or part filled)

Object	balance reading in air	balance reading in water
stone		
bottle		
brick		

This experiment will help you to answer some of the questions that should arise from your last investigation.

| object | SCALE READINGS | | | bucket of water plus object immersed |
	bucket of water	object in air	object in water	

Some More Advanced Investigations

Continue to investigate with a special piece of equipment called an overflow can.

1 Look at this apparatus :—
Where must the water level be to start ?

What will happen when you immerse an object ?

What must you make sure is in position before doing this ?

2 Use this apparatus :—

Immerse the object and notice what happens.

Now that you understand its use you can make careful observations and measurements.

This experiment involves balancing forces. Our force measurements should be in NEWTONS.

1 What is the volume of the object (in cm³)?

2 What is its mass (in grams)?

Note – These two measurements must be the same in air and water

3 What is the downward pull of the object in air (in newtons)?

4 By how much does this force change when you immerse the object in the water ?

5 What volume of water is spilled over ?

6 What is the mass of this water ?

7 What is the downward pull of this water? (First you need to find the downward pull of the container.)

8 Can you find any relationships between these measurements? (Look particularly at the force measurements.)

CARGO
60,146 tons

CARGO
60,146 tons

32
31
30
29
28
27
26
25
24
23
22

32
31
30
29
28
27
26
25
24
22

SALT
WATER and FRESH
WATER

TF
F
L O R T
S
W
WNA

PLIMSOLL

SECTION
TWO

Salt water and floating

How does salty water affect floating things ?

1 Prepare some salty water (try 50g in a jam jar of water).

2 Make a floater (it will help to have a method of marking these).
Some suggestions :

A thin strip of balsa wood with drawing pin to load the end.
Mark direct onto wood.

Small test-tube, loaded with sand or lead shot.
Marked strip of paper inside.

Cigar container loaded with sand or lead shot.
Scratch marks on side.

Straw sealed with plasticine, loaded with lead shot.
Mark direct with fibre-tip pen.

Mark all types of floater in $\frac{1}{2}$ centimetres to record positions.

3 Try your floater in the salty water.
4 Try your floater in the tap water.
5 Try your floater in different salt concentrations :

    a 100 g to a jam jar.
    b 25 g to a jam jar.
    c 10 g to a jam jar.

Can you find out how salty sea-water is, or let a friend or teacher mix an unknown concentration ?

Fibre-tip mark showing depth of the floater when in tap water.

Record for each kind of water.

6 You could try your floater in other liquids: methylated spirits, milk, paraffin, vinegar, cooking oil, turpentine, sugar solutions.
Record the floating line again for your new liquids — compare with water. Can you discover how and where such observations are used?

Will a ship float higher or lower in salt water than in fresh water?

How has the salt changed the way a ship floats?

This young lady is floating in the Dead Sea.
It seems very easy doesn't it?

Does your research help you to explain why?

The following experiment will help you to see a difference. You will need: 4 jam jars, 2 pieces of stiff card and a large bowl, sink or tray. Fill two jars with plain tap water and the other two with coloured salty water (100 g to the jar).
Place as shown, working over sink or bowl.

coloured salty water

card

plain water

plain water

coloured salty water

To fill inverted jam jars:—

1 Fill to top of jar.

2 Cover with stiff card

3 Turn over jar with card firmly in position.

4 Lower carefully onto the other jar.

5 Remove cards when in position.
Which is denser — how does this affect floating?

Some book research

You will need to use your library books to help you with your investigations. Some interesting subjects for investigation are shown below. Try and find answers to the following questions :—

SHIPS' MARKINGS

1 Why are they there ?
2 What do they mean ?
3 What do the letters stand for?
4 Who was responsible for this law of the sea ?
5 Why was his reform so necessary ?

SHIP TERMS

6 What do sailors mean by the trim of the ship ?

7 What is ballast ?
8 What is meant by 'draught', 'keel', 'freeboard' ?
9 What is deadweight tonnage ?
10 What other kinds of tonnage are there ?

Draw a picture of a ship. Put these 'shipwords' in the appropriate place :

a) bridge g) forecastle
b) hull h) amidships
c) aft i) starboard
d) stern j) foremast
e) keel k) funnel
f) port l) poop

Two examples of the ship's marking investigated above.
left: on the side of the 'Cutty Sark', right: on a modern ship.

The SHAPE of SHIPS — *and why*

CUTTER

RAM

STRAIGHT

CLIPPER

RAKED

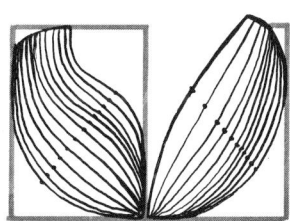

SECTION THREE

Looking at Shapes

Why are these boats shaped as they are? Which shape will travel through water the fastest? Which shape will travel with least turbulence?

speed boat

punt

kayak

racing catamaran

oil tanker

elevation of hull

ice breaker

Some suggestions for investigating shape, speed and motion:-

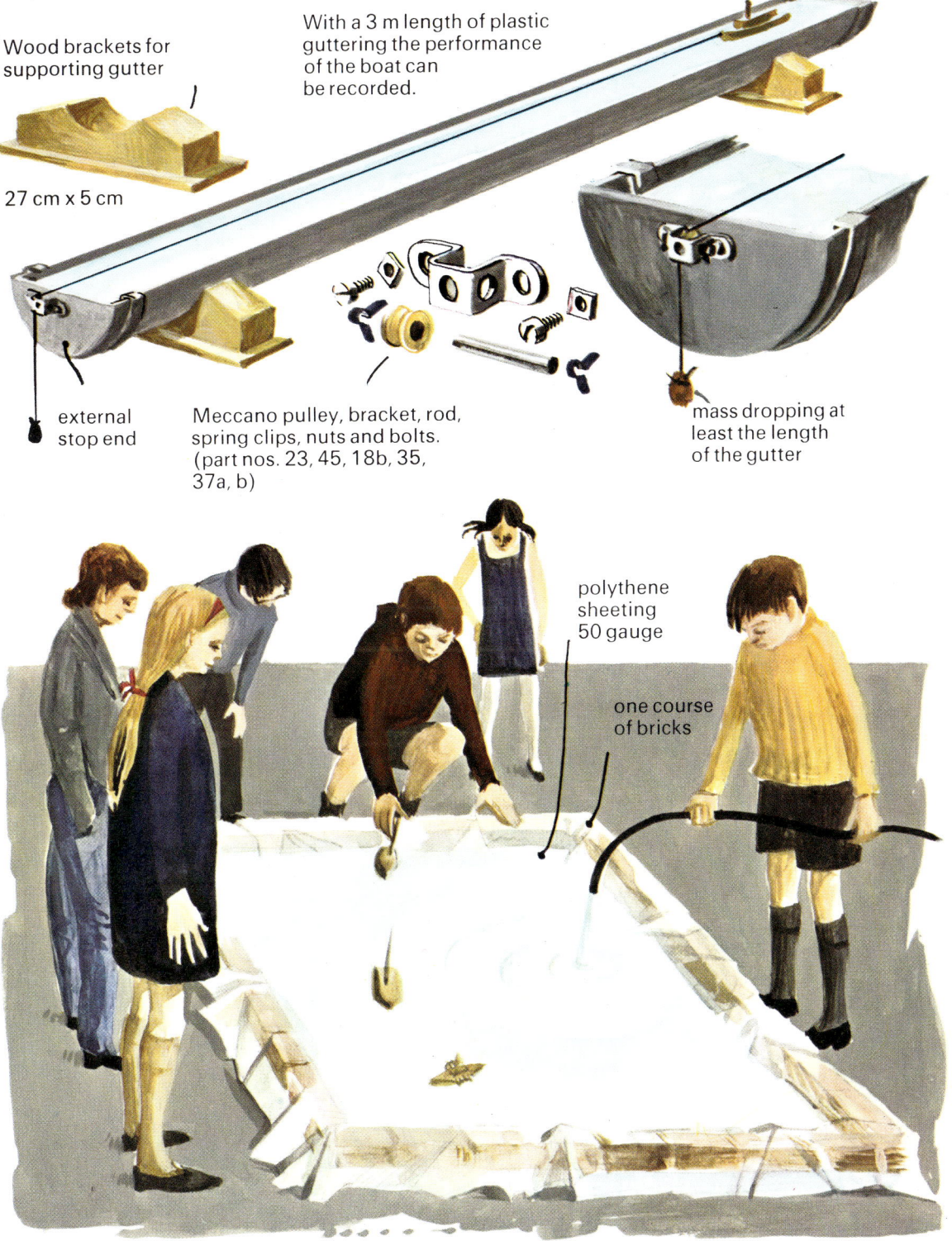

Wood brackets for supporting gutter

27 cm x 5 cm

With a 3 m length of plastic guttering the performance of the boat can be recorded.

external stop end

Meccano pulley, bracket, rod, spring clips, nuts and bolts. (part nos. 23, 45, 18b, 35, 37a, b)

mass dropping at least the length of the gutter

polythene sheeting 50 gauge

one course of bricks

Children filling a playground pool, made with bricks and covered in plastic sheeting (or use an inflatable paddling pool) for further research.

Shapes we might investigate:

What factors could influence performance?

1 Increasing the length

— small nails for attaching pulling threads

4 Shaping the hull.

2 Increasing the width.

5 Smoothing the hull.

sandpaper block

3 Increasing the depth.

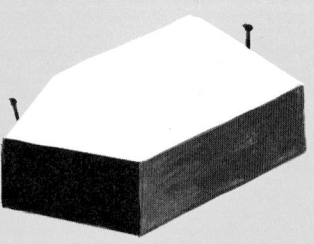

6 Painting the hull.

Ink dropped in behind the boat could show the turbulence.
Does increase in speed affect turbulence ?
How are designs for real ships tested ?

How are ships BUILT?

SECTION FOUR

23

Model Making

Children making models

You can build models that will tell the story of shipbuilding :—
The 'Santa Maria' and sailing ships of the 15th century were built to the ratio 1 : 2 : 3 – beam, keel and overall length.

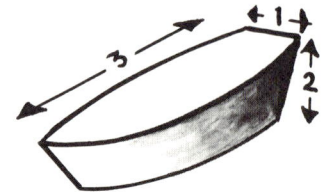

1 Make a model to this ratio of 1 :2 :3.
2 Compare this model in your test tank with models of widely varying ratios.
3 Use your reference books and scale models you have made to find which ships were built to this ratio and if new materials and ship-building methods have changed this early 'rule of thumb'.

These models were made by children between ages 9-11

Plans of some of the ships are shown inside covers.

The origins of ships :
log, dug-out boat, log-raft

A Grecian
galley
(4th century B.C.)

A Roman merchant
ship (2nd century B.C.)

A Viking warship
(9th century A.D.)

Children's models

above : 13th century English warship
— 'castle' type.
left : Tudor warship
below : Early steam boat – early
19th century.

The Great Eastern (1858)

Four examples of modern passenger
liners – models by 9 and 10 year old boys.

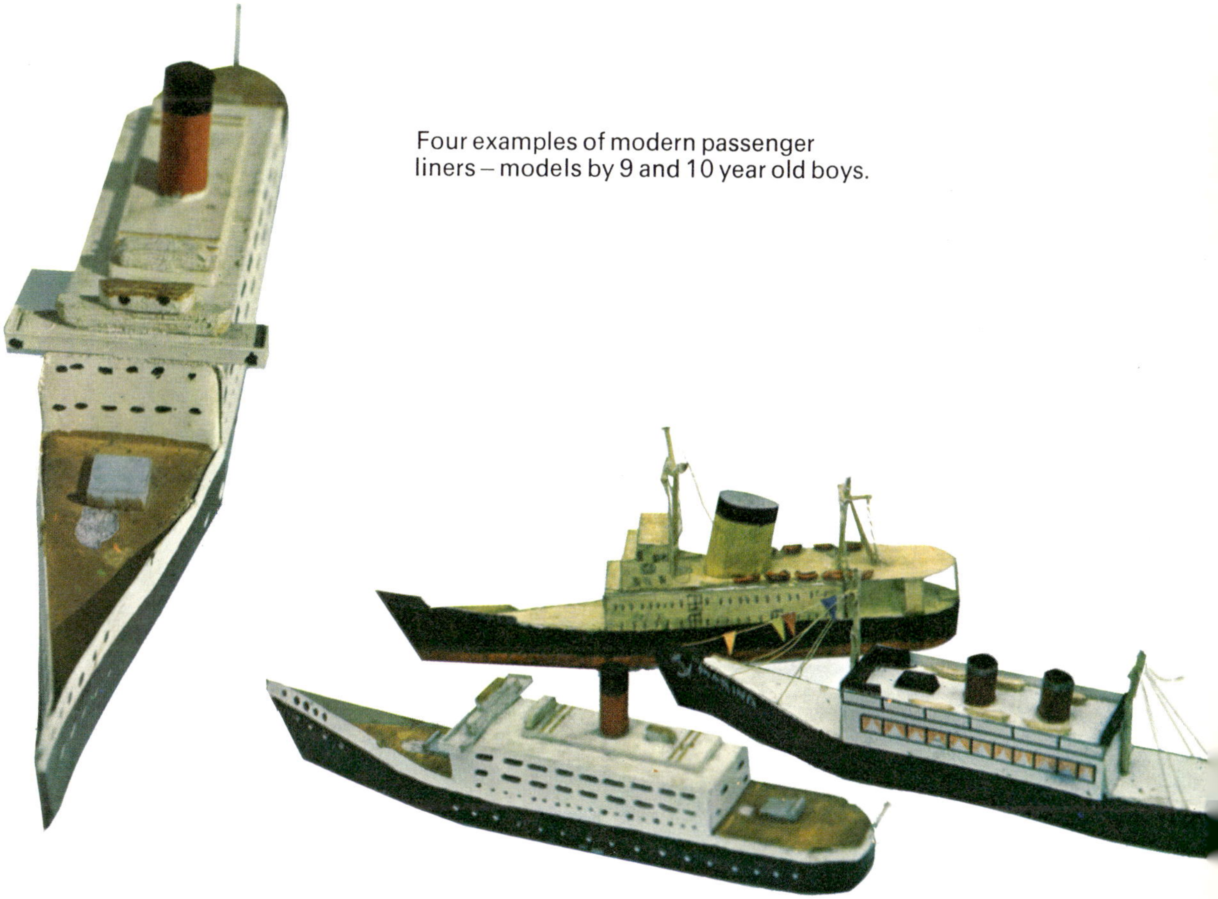

Fixing and Fastening

How was the planking of the Santa Maria and such early wooden ships fastened?
What can you find out about 'clinker' built and 'carvel' built?
How are the steel plates of modern ships fastened?

RIVETING AND WELDING

Investigate types of rivet fastenings.
1 Make a butt joint and a lap joint using hardboard and paper fasteners.

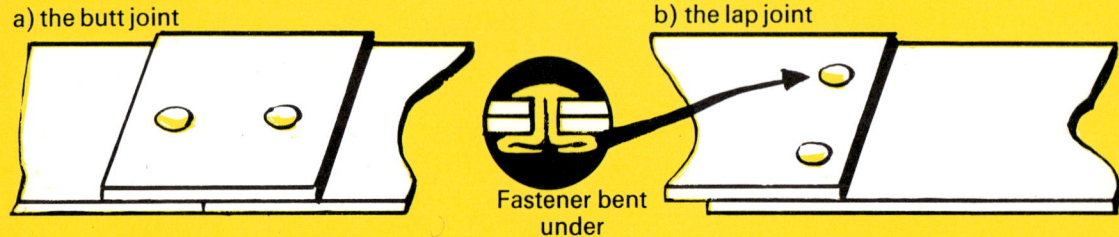

a) the butt joint

b) the lap joint

Fastener bent under

2 Try fastening metal, tin sheet, etc. with solder. Welding the plates in a modern shipyard involves melting steel. You cannot do this in class but you could experiment and test soldered joints and compare a similar joint using a modern adhesive like 'Araldite'.

3 Does increasing the number of rivets per joint make it stronger?

A simple test for your rivets:

measuring and recording

The Electro-Magnetic Crane

An Electro-Magnetic crane is often used in shipyards and steelworks to move heavy steel plates.

A 5-magnet crane transporting steel plate of approximately 4 tonnes at the BSC Hartlepool Works.

Left: Children using a model of an Electro-Magnetic Crane. If you have studied Electricity and Electro-Magnetism you may like to make a similar type of model.

Below: a simple switch to control the crane.

stranded wire

strip steel

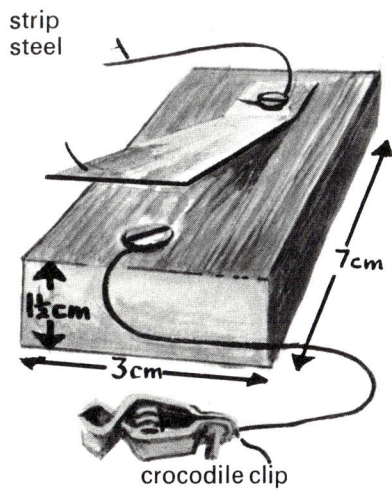

crocodile clip

Rotting and Rusting

Wood rots, iron rusts – so how were these materials protected, and how are they protected today?

What are teredos? What protection was tried? What was careening? How was it done?

1 Examine some rotting wood; what changes have occurred? How many different ways can you find of preserving wood?

Rusting is a problem whenever iron is used.

Research into Rusting and its Prevention

Useful forms of iron to experiment with include: steel wool (remove grease with a solvent such as Thawpit), iron nails and iron filings.

The following diagrams will give you some suggested conditions to help with your research:

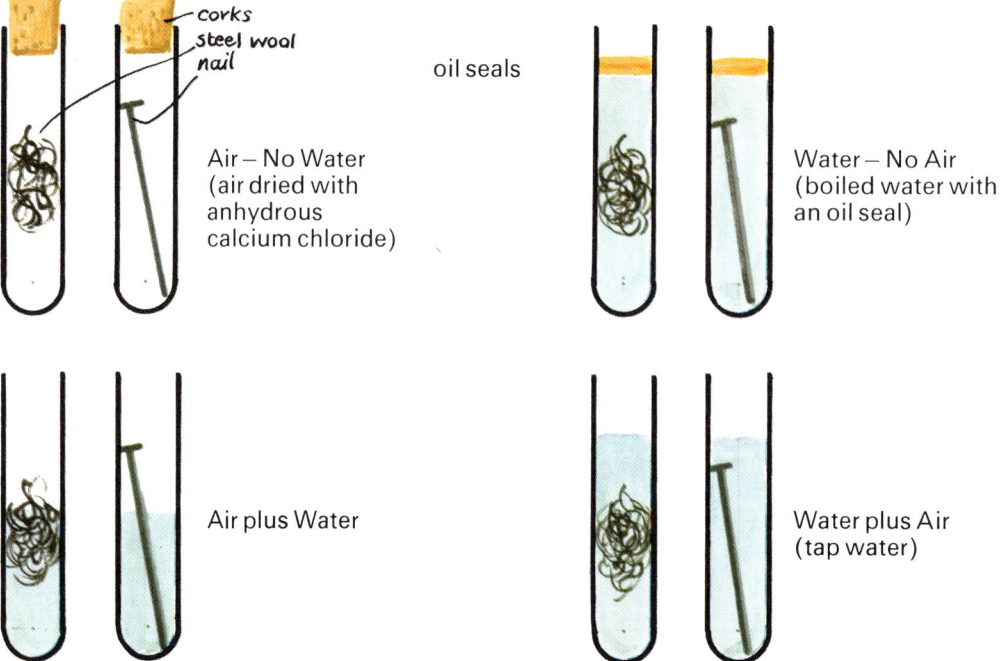

corks
steel wool
nail

Air – No Water (air dried with anhydrous calcium chloride)

oil seals

Water – No Air (boiled water with an oil seal)

Air plus Water

Water plus Air (tap water)

You will need to make careful recordings of :– a) the date the experiment was set up b) the daily appearance. c) do this for 12 to 14 days.

3 Do any other metals corrode (rust) under these conditions? Try copper, lead, zinc, aluminium and other metals.

4 As ships sail in the sea, it would be interesting to repeat these experiments using salt water.

What ways can you find of protecting iron?

Try protected samples under the conditions you have found that cause rusting.

Some suggestions for research :–

1 Nail half-submerged in water. Where does rusting first occur?

2 Scratched piece of tinplate. Also try scratching tin cans and leaving in various places.

3 Galvanised nail or piece of galvanised steel – scratched.

4 Painted nail – scratched

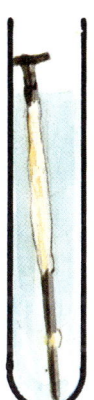

5 Partially greased nail.

6 Copper plated nail (scratched) – you can copper-plate by placing in copper sulphate solution.

7 Do germs and bacteria affect rusting? Experiment with germicidals such as TCP & Dettol.

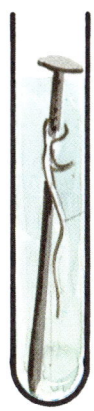

8 Nail with magnesium ribbon attached.

In which ways can ships be protected?
Can you think of any other ways that might work?

Slipping and Sliding

How are ships launched?
How are they held until ready?
How do they slip into the sea?
How is the slipway made slippery?
Sometimes we want two surfaces to grip –
sometimes to slip!

Try making a model slipway for your research. A thin edging would contain any lubricants you use.

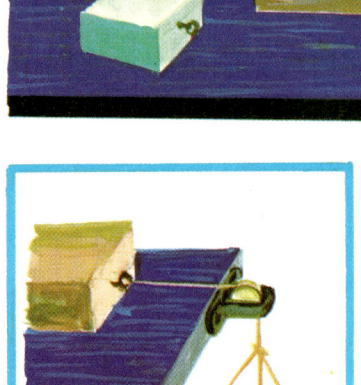

You can easily find this force by hanging the pan on a newton balance.

Another way using scales and masses

Record your investigations like this :—

Find the best slipping situation.

SURFACES UNDER INVESTIGATION

1 To start, you could try plastic on plastic.
2 Try lubricating with water, oils and greases, talcum powder, graphite and polystyrene beads.
3 You could try varying the surfaces in contact – the area in contact – the pressure by placing different masses on top of the block.
4 As an alternative starting force you could try varying the slope of your 'slipway'.

You will find it best to record your results immediately to avoid confusion. You can then think of ways of showing the results to others.

surfaces in contact	area of contact (cm²)	lubricant	moving forces (newtons)

Can
Boats
**TOPPLE
OVER?**

Balancing Flat Shapes

To investigate this problem of stability we need first to find out how things balance, and topple over.

a) with your hand

b) with a pencil

1 Find the balance point of a ruler. Try again with one end loaded with plasticine.

2 Try to find the balance point by trial and error using your pencil point.

3 Now find the balance point of some flat shapes.

Here are some suggestions for you to cut from card :—

You may find it difficult to work with shapes smaller than 20 cm. across. A more efficient way is to use a plumb line made from a piece of thread and a small washer.

Finding the Balance Point

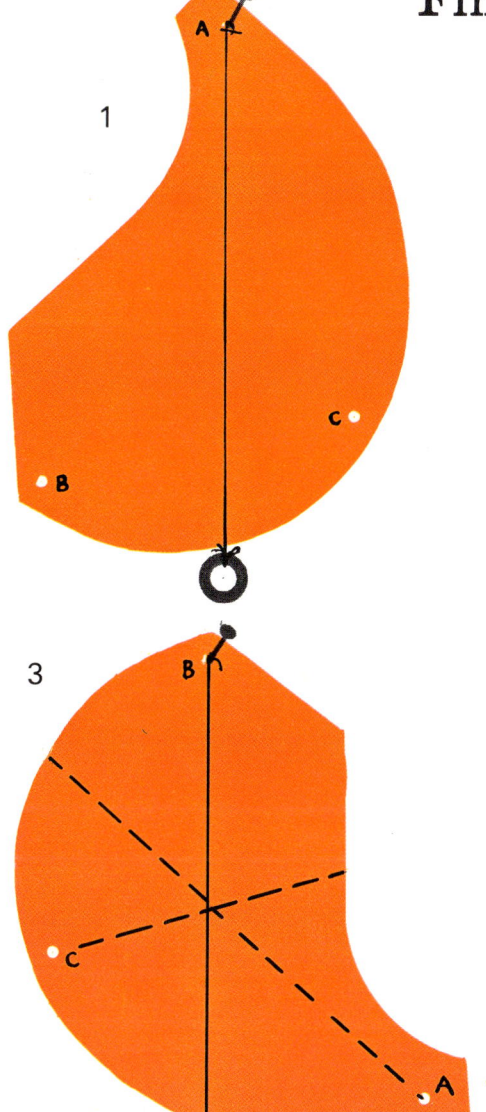

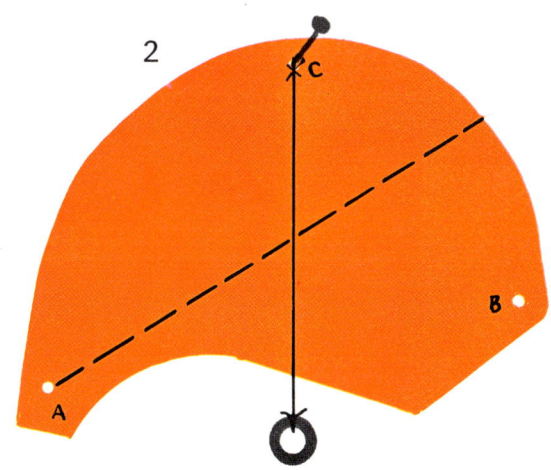

1 Cut a piece of cardboard of an irregular shape.

2 Pierce 3 holes anywhere near the edges of the shape.

3 Pivot on a nail or pin and hang a plumb line of thin string with a washer.

4 Mark the position from point A and then repeat as shown here on points B and C.

5 You can check that this is the balance point.

This balance point is called the CENTRE OF GRAVITY

How does the centre of gravity alter if you change the mass distribution? (You could use plasticine or tape on a washer or coin.)

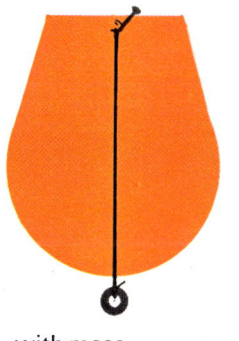

with mass

with mass on right

with mass on left

repeat pivoting from other points, as in the example shown above

The Centre of Gravity

How does the Centre of Gravity apply to a ship shape?

1 Cut out this white ship shape from a piece of stiff card (you may be able to trace this one).

Find the centre of gravity using a plumb line as before.

2 Now try tilting tests using a test rig as shown.

Observe carefully the relationship of plumb line to base when the 'ship' becomes unstable and topples over.

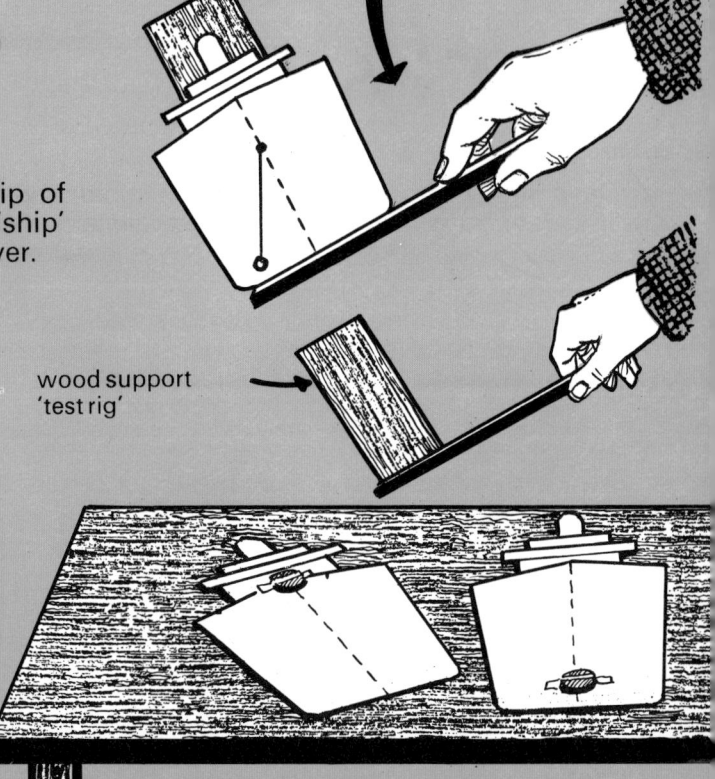

wood support 'test rig'

3 What difference does the addition of 'ballast' or 'deck-cargo' make?

Try taping a washer or coin to the top and then the bottom of your shape:

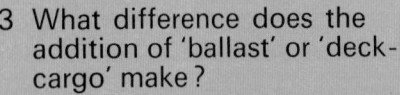

Balancing Solids

You could continue your investigation with solid objects.

1 Use a rectangular block of wood with a plumb-line from point 'A'. Try tilting and notice where the plumb-line is at the point of toppling . . .

Experiment with a toy bus-load on top and load under (using lead or nails).

Try tilting with the plumb–line attached to a corner.

2 Are all bottles equally stable? How does changing the volume of liquid affect stability (note 'C' half-filled)?

medicine

INK

Scent

a   b   c

INK

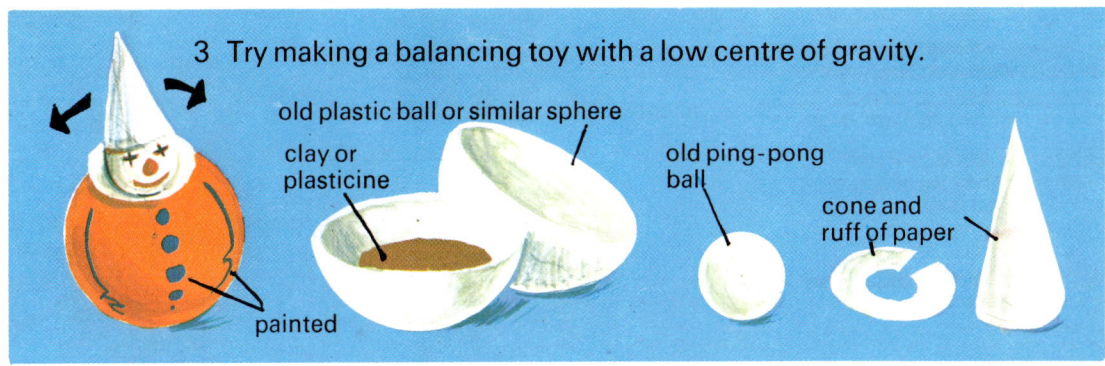

3 Try making a balancing toy with a low centre of gravity.

old plastic ball or similar sphere

clay or plasticine

old ping-pong ball

cone and ruff of paper

painted

4 Try further ideas with things that have a low centre of gravity.

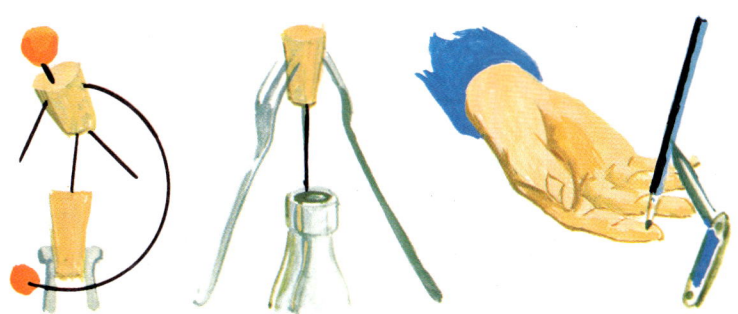

Note where the centre of gravity and point of support is in each example.

Stability – 1

GRAVITY   **G**

The stability of a ship lies in its tendency to right itself and depends on two forces :—

in still water

BUOYANCY   **B**

In rough water which ship is unstable ?

in rough water

A ship called the Vasa came to grief in 1628 – can you find out why ?
Shipbuilders have to use complicated mathematics to ensure their ships remain stable in any sea. If you wish to find out more about this, look up Pierre Bouger and his work on the 'Metacentre'.

A model to investigate stability

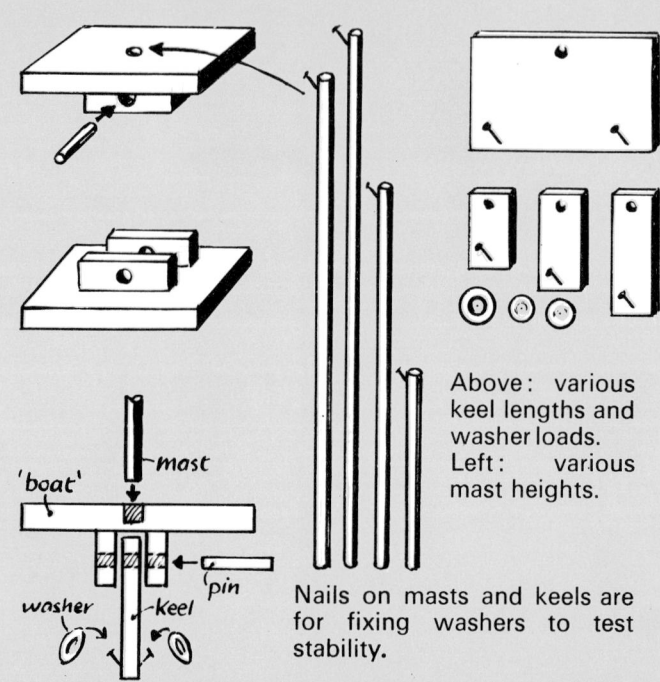

If you would like to make a representative ship you could investigate the effect on stability of various heights of masts and various keels.

How does the height of the mast affect stability ?

Does loading the mast increase or decrease stability ?

When you have made your ship unstable can you make it more stable by altering its keel ?

Here are some diagrams that may help you :—

Above: various keel lengths and washer loads.
Left: various mast heights.

Nails on masts and keels are for fixing washers to test stability.

Stability – 2

1 Does length improve stability?

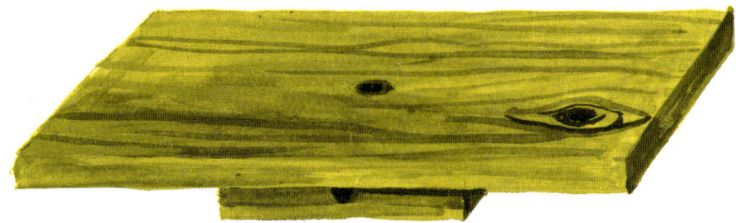

What would happen if you made your 'boat' twice as long? (2 x L)

a dug-out canoe

2 Does width improve stability?

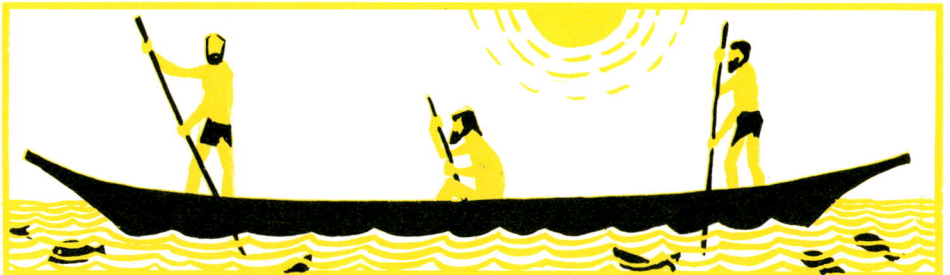

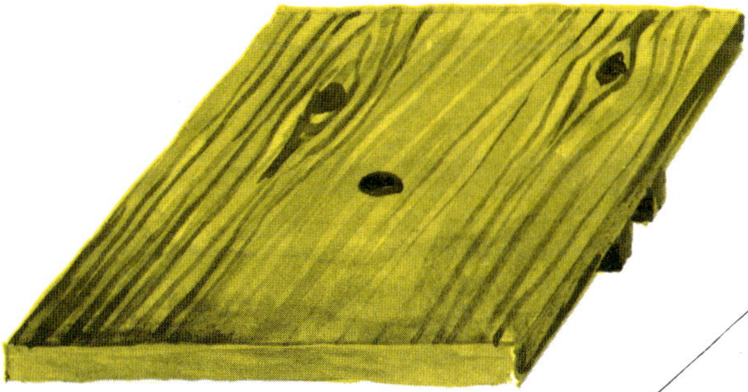

Would the stability be improved if you made your 'boat' twice as wide? (2 x B)

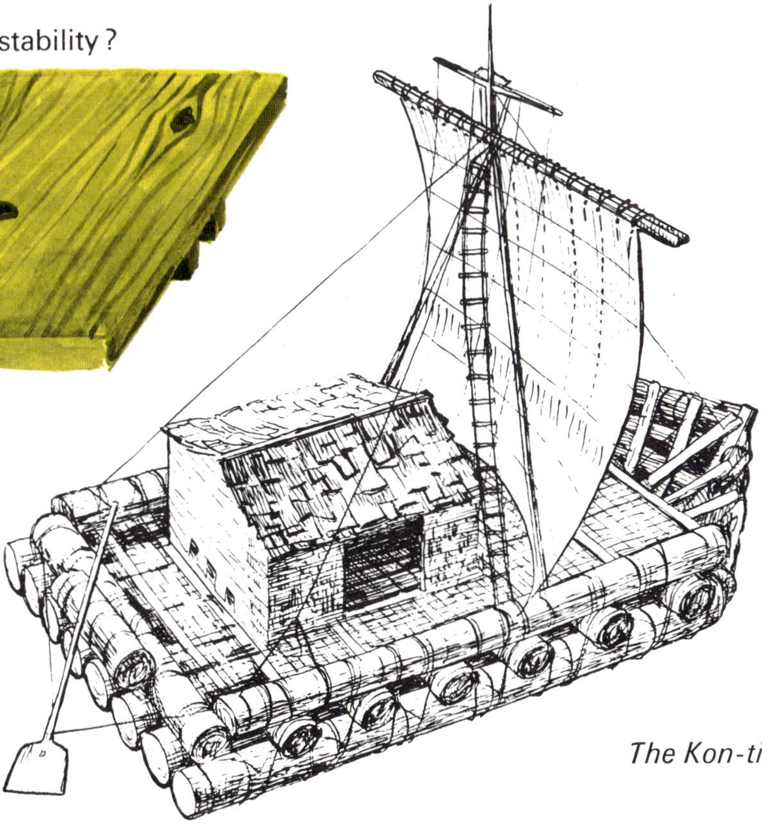

The Kon-tiki

Stability – 3

3 Make your boat twice as deep (2 x D).

Does draft improve stability ?

A cargo vessel

---

4 Does a twin hull improve stability ?

A racing catamaran

Try joining two 'boats' as shown.

How do your findings relate to the actual boat shapes ?

How are Ships MOVED?

SECTION SIX

Methods of Moving

How many ways of propelling ships can you find?

How many ways of propelling your toy ships can you devise?

1 The Oxford and Cambridge Boat Race

2 Yacht

3 Two Hovercraft SRN2 at their berth at Dover

4 'Sirius' paddle steamer – 1837

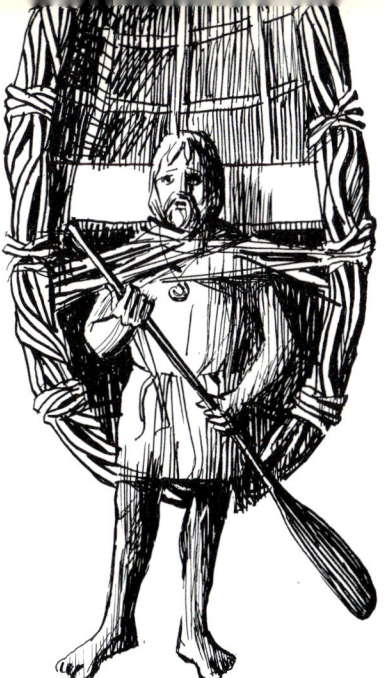

Arms, Paddle, Oars

Probably man's first means of moving his early boat would have been with his arms and then with the paddle.

Is the paddle still in use today?

The oar is a development from the paddle and an interesting use of a lever that we can investigate.

In Roman times the Celts were using a boat like this. It is called a coracle and is made from branches covered with hide.

The coracle is still used. Where? What is it made from today?

5 mm. dowel

Here is a simple test-rig which may help to explain what is happening when you row a boat.

As you move the oar to move the 'boat' along think about these questions:

1 What is the load that is being moved?

2 Where was your force applied?

3 Which part of the oar is the turning point?

4 Whenever we use a lever like this there is always :—

a load
an effort
a turning point or pivot

A   B   C

5 Where on the above diagram of an oar will you find a load, an effort and a turning point?

Propulsion – 1 Oars

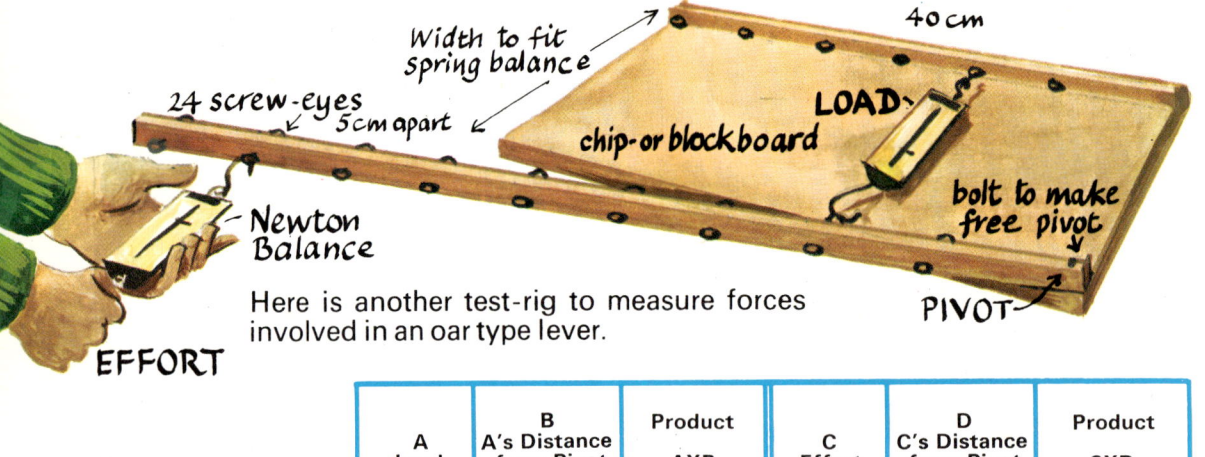

24 screw-eyes
5 cm apart

Width to fit
spring balance

40 cm

LOAD

chip- or blockboard

bolt to make
free pivot

PIVOT

Newton
Balance

EFFORT

Here is another test-rig to measure forces involved in an oar type lever.

You can collect the results in a table like this:

A Load	B A's Distance from Pivot	Product AXB	C Effort	D C's Distance from Pivot	Product CXD

The oar is one particular type of lever – can you make a list of other types of lever? You could look in and around your school and home. It will help you to understand how these inventions can make our work easier if you draw a simple picture of each, and label the LOAD, EFFORT and PIVOT (another word for pivot is FULCRUM).

Here is one labelled example:

EFFORT (by hand)
LOAD (grip between tin and lid)
PIVOT (edge of paint tin)
PAINT

Here are some more levers for you to label :–

double levers

biceps muscle

How many different levers can you label in this way?

Propulsion — 2 Sail

Man soon found that wind could be used to propel his boats.

1 Model some of the various types of sail and try them out. You could do this with model boats on a pool.

block balsa
What difference does a keel make?

2 Here is another suggestion for trying out the sail indoors, using a fan or hair dryer.

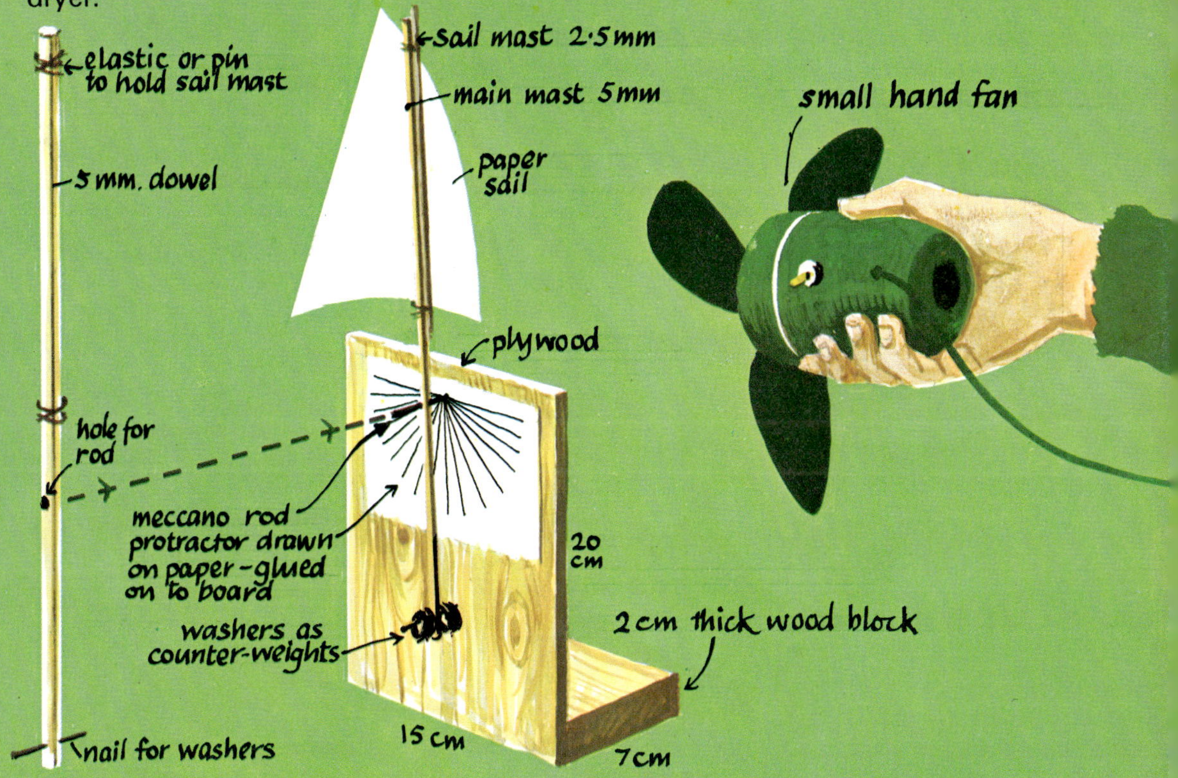

elastic or pin to hold sail mast

5 mm. dowel

hole for rod

meccano rod
protractor drawn on paper – glued on to board

washers as counter-weights

nail for washers

sail mast 2.5mm

main mast 5mm

paper sail

plywood

20 cm

15 cm

2 cm thick wood block

7 cm

small hand fan

Here are some sail types for you to try :—

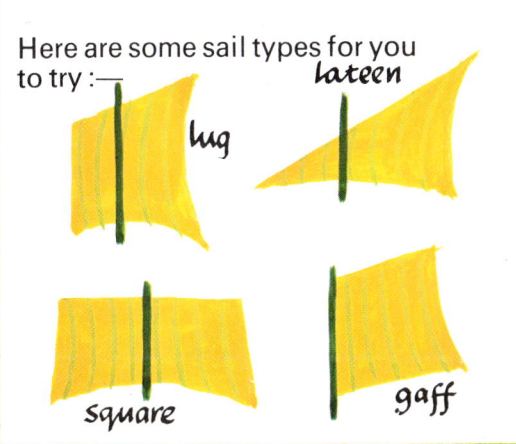

lug

lateen

square

gaff

Try blowing from different quarters.

1 Try different shapes with the same area.
2 Try different shapes with the same height.
3 Try different areas using the same shape.
4 Does curving your sail make any difference?

Propulsion – 2 Sail (continued)

This wind will take this ship . . .

. . . in this direction

what happens if the wind is blowing . . .

. . . and the ship needs to sail . . .

. . . in this direction, or this, or this ?

How must the sails be set ? If you would like to find more about how sailing craft are manoeuvred do try to talk with a yachtsman.

A Tahitian outrigger

It might be fun to model a special type of boat used by Polynesian and Tahitian sailors.

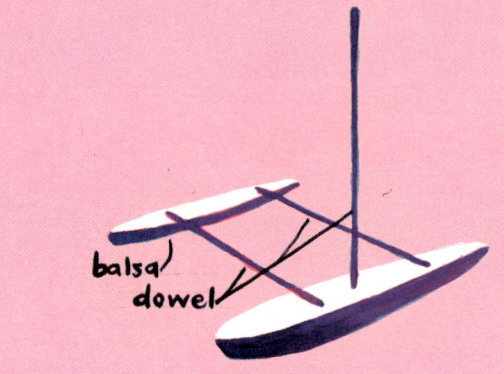

balsa dowel

Which side should the outrigger be . . . this . . . or this ?

Propulsion – 3 Steam

At the close of the 18th century a new form of power was found – steam.

If you collect the items shown in the illustration below you can make a steam turbine :—

Steam engine—1843

tin and lid with hole punched in centre

aluminium sheet or tin for turbine blades

washers

meccano strips

tin snips

tape

meccano parts

twist blades

bend→ cut

7cm.dia.

Try various designs of turbine

The completed turbine

tape→ holds meccano strips

gas

How was the new power used to drive the ships?

You could find out about how the Paddle wheel and propeller were developed in the early steam-ships.

Some of the most famous of these were :

Symington's CHARLOTTE DUNDAS, Fulton's 'CLERMONT', Bell's COMET,

Brunel's THE GREAT BRITAIN.

The Ships Gallery of the Science Museum would be a splendid place for this investigation. You could investigate propulsion using the paddle wheel and propeller by modelling boats with rubber bands.

Propulsion – 4 Paddle and Propeller

paddle propulsion

propeller propulsion

1 Make a model paddle boat.

tacks, elastic band and two pieces of balsa

2 Make a model of a propeller boat using soft wood, angle bracket, hook and bead. Lubricate between bead and angle.

glass bead   underside

  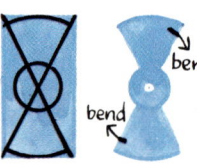

stages in making the propeller using sheet tin or aluminium

Which is the better method of drive — propeller or paddle. ?

In 1845 a tug-of-war between propeller and paddle took place to find out the answer to the above question. How did they make sure the trial was a fair one ?

Can you restage the trial with models ?

Propulsion — 5

What other ways of propelling a model boat can you find or invent?

AIR SCREW

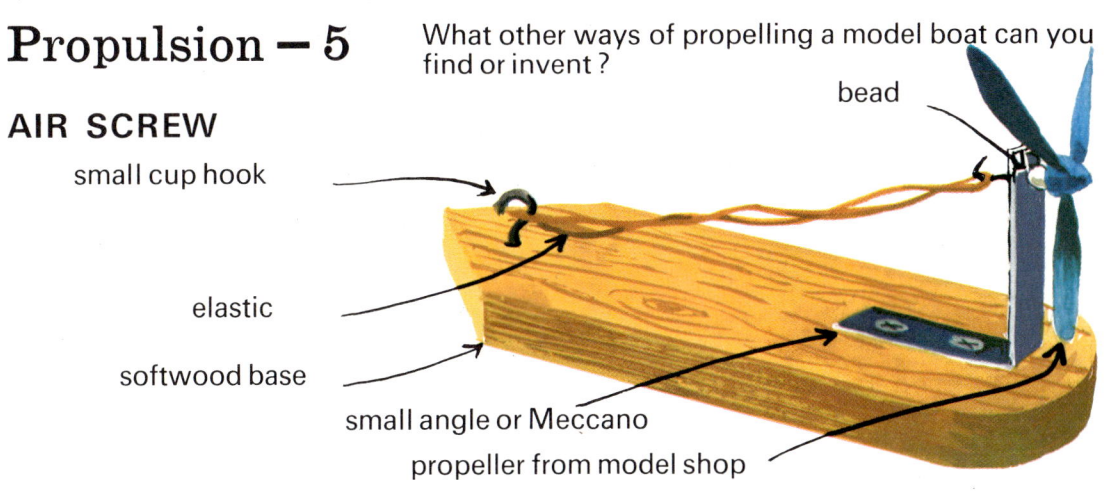

small cup hook

bead

elastic

softwood base

small angle or Meccano

propeller from model shop

JET-PROPELLED

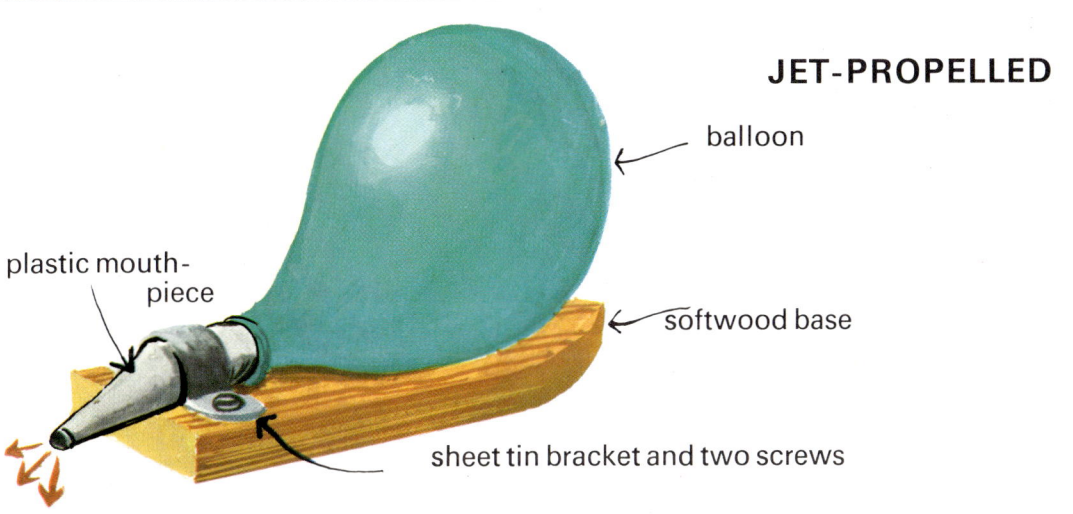

balloon

plastic mouth-piece

softwood base

sheet tin bracket and two screws

CLOCKWORK MODELS

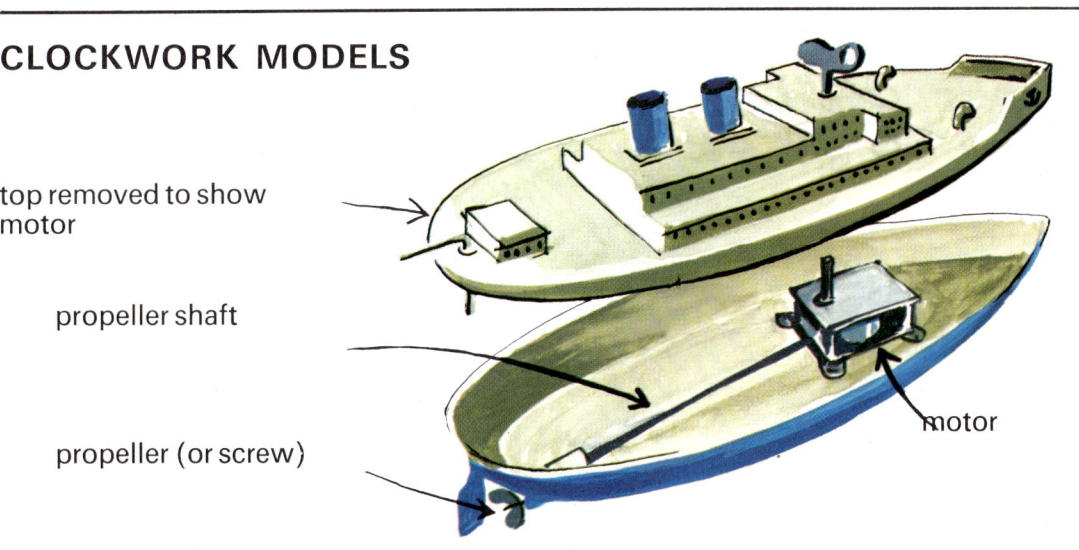

top removed to show motor

propeller shaft

propeller (or screw)

motor

Propulsion – 6

STEAM JET

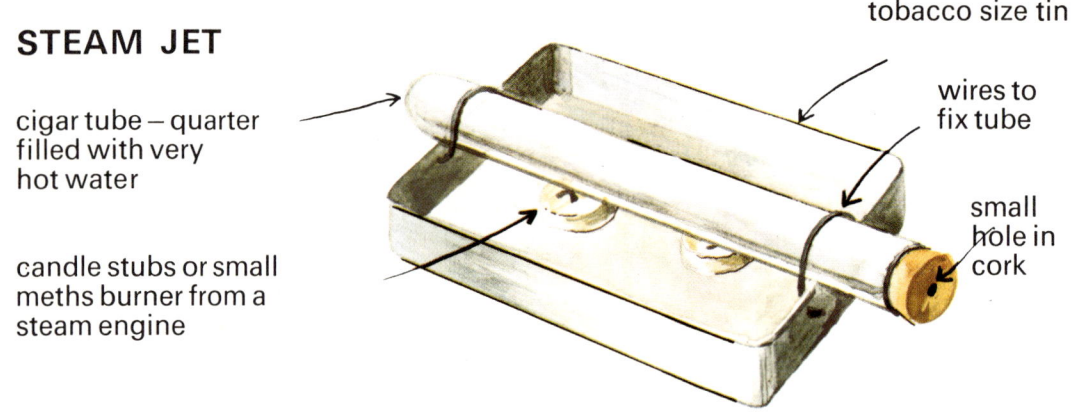

cigar tube – quarter filled with very hot water

candle stubs or small meths burner from a steam engine

tobacco size tin

wires to fix tube

small hole in cork

'CAMPHOR' or 'OIL' BOATS

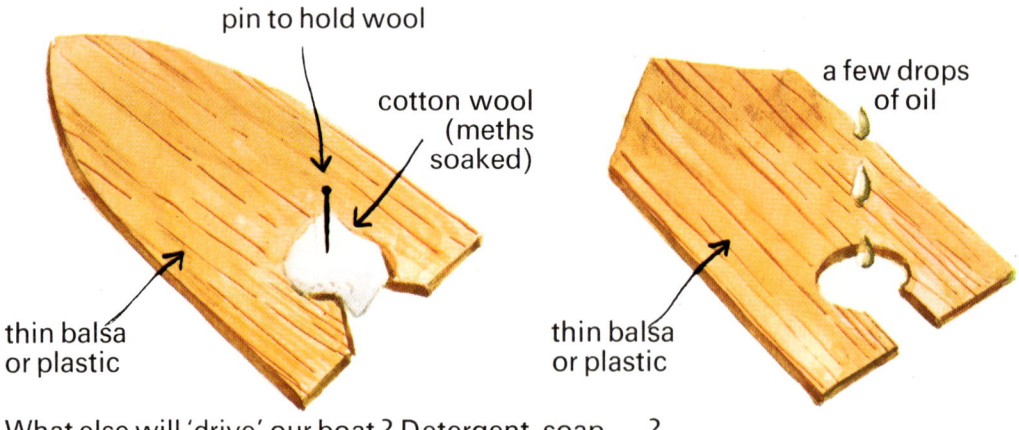

pin to hold wool

cotton wool (meths soaked)

thin balsa or plastic

a few drops of oil

thin balsa or plastic

What else will 'drive' our boat? Detergent, soap . . . ?

PADDLE BOAT

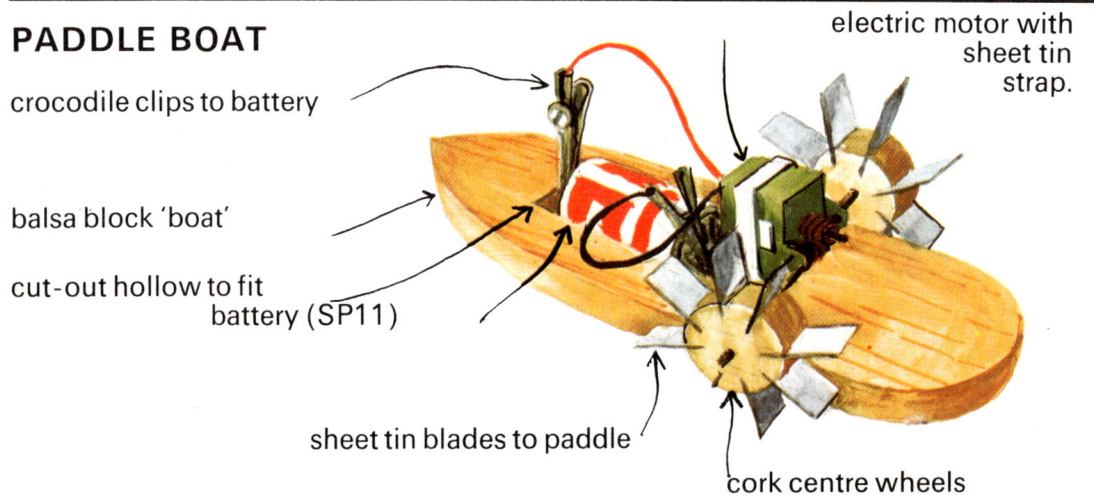

crocodile clips to battery

balsa block 'boat'

cut-out hollow to fit battery (SP11)

electric motor with sheet tin strap.

sheet tin blades to paddle

cork centre wheels

Anchors are raised
Sails are hoisted —
HOW?

Section Seven

Raising Anchor

What are parts 'A' and 'B' for in the photograph?

Can we measure how effective an anchor is? You could research using one of your model boats.

What other types of anchor have been used? Can you find out the names and uses of the types shown here?

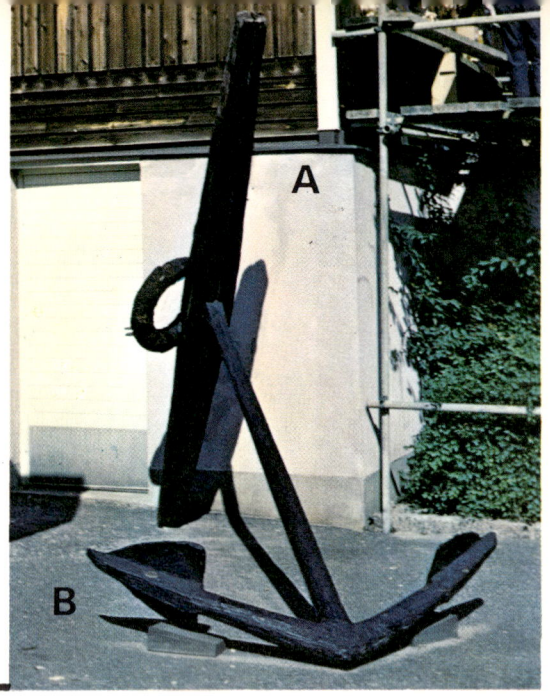

1 Can you make a model of a capstan similar to the one below and find out how it was used to raise the anchor?

Here are some sailors raising anchor with an early capstan.

2 What happens when we let go? Can you improve your model to prevent 'slipping'?

3 The men above are turning the capstan, but how does a large modern ship raise anchor?

4 Can you invent a method of raising an 'anchor' using a toy motor? (Toy electric motors are quite inexpensive and easily obtained).

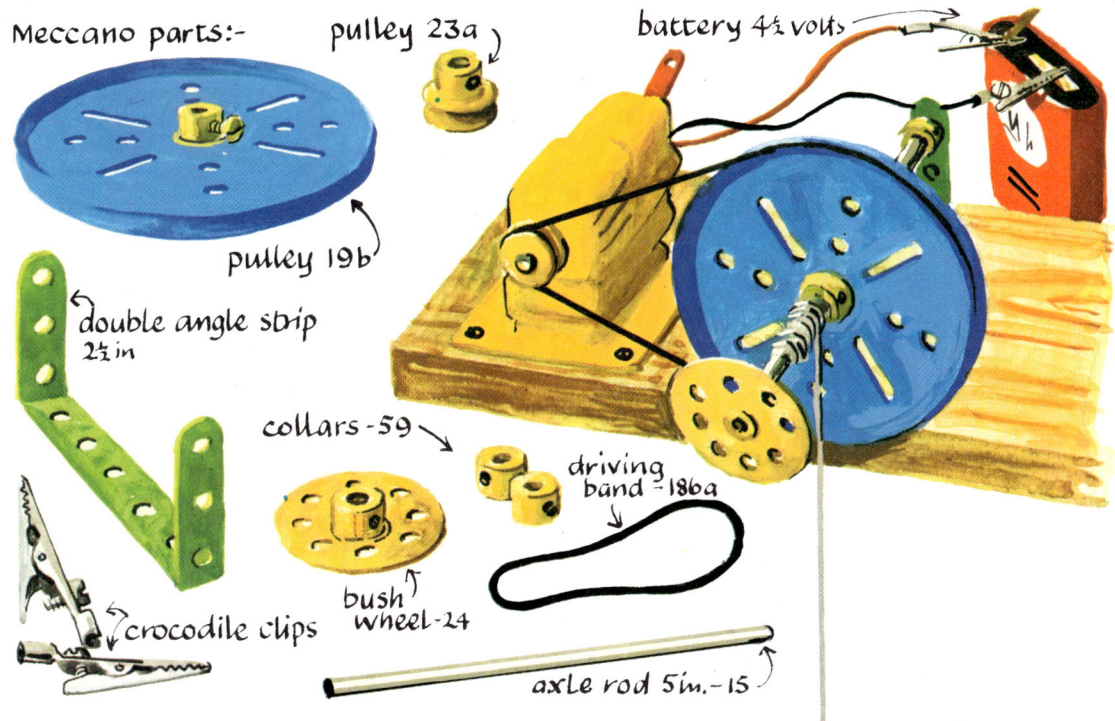

Meccano parts:-

pulley 23a

battery 4½ volts

pulley 19b

double angle strip 2½ in

collars - 59

driving band - 186a

crocodile clips

bush wheel - 24

axle rod 5in. - 15

A suggested 'Test Bed'

Build a mechanical capstan as shown here. Your electric motor, a Meccano type as illustrated above, using a 4½ volt battery could provide the energy.

1 How much 'work' is done lifting the mass ?
2 How powerful is the motor ?
3 What different ways of connecting the drive can be found ?
4 What difference would gears make ?

You might compare different kinds of motor and try different batteries within the voltage range of the motor.

Meccano motor type 11055 is a 4½ volt reversible D.C. Motor.
Type 11057 is a 3-12 volt motor with a 6 ratio gearbox.

Compare an electric motor with other types :—

a) clockwork
b) steam
c) hot air
d) elastic
e) model aircraft engine

wire hook

loaded plastic pot or bucket

Pulleys

It is very difficult to move heavy sails. Pulleys are used to make this work easier. You could investigate how different pulley systems help.

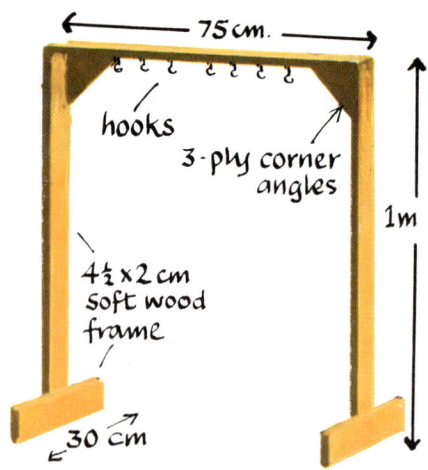

75 cm.

hooks

3-ply corner angles

4½ x 2 cm soft wood frame

1m

30 cm

You could use a doorway or a simple test-rig that could be constructed as shown here :—

Below : the test-rig in use.

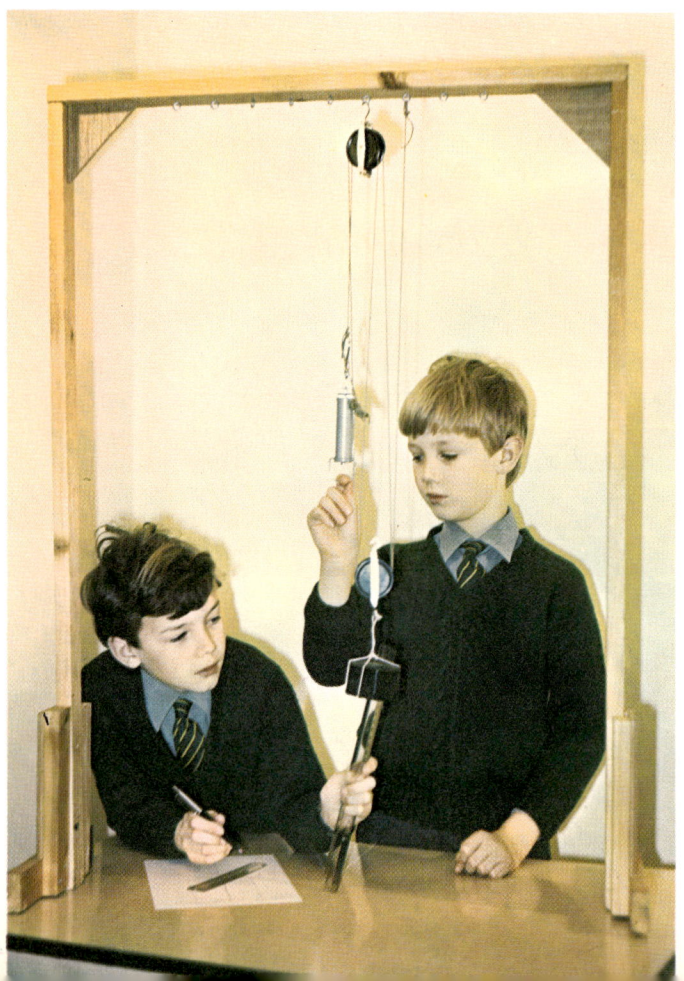

1 2 3 4

Four pulley arrangements

The 'Victory' used over 1,000 pulley systems aloft.

Can you devise a)   3 pulleys in a system?
         b)   4 pulleys in a system?
         c)   ? pulleys in a system?

Record your results for the various arrangements of pulleys. You should think about your choice of units for this investigation. Remember that you are experimenting with two forces. First the force you are exerting to pull up the sail and secondly the force acting downward resulting from the mass of the 'sail' which is being raised. We measure forces in NEWTONS.
You can measure how many newtons we need to pull up the 'sail mass', by pulling with a spring balance calibrated in newtons.

You can find the downward force exerted by the 'sail' by hanging the mass on the newton balance. You can then record your results on a table :—

downward pull due to mass of sail	your pull	distance moved by sail	distance moved by your hand
Newtons	Newtons	Metres	Metres

1 Which system felt the easiest to use?

2 Which system would be impractical for sail raising?

3 Is there any connection between ease of pull and distance pulled?

4 Can you think of any difficulties there would be in using a 20 pulley system?

Work and 'Joules'

You must have felt that in pulling up sails and raising anchors that you have been doing WORK.

These two pictures show work being done.

Work is a term which is sometimes used rather loosely in everyday conversation.
In science WORK always has a precise meaning.

> ## Force X Distance Moved = Work

The bigger the force used the more WORK you do.
The further you pull the more WORK you do.

Work is measured in JOULES :

> ## ∴ Newtons X Metres = JOULES

1 How many JOULES were needed to raise your sail in each case? Check with your investigations on page 55.

2 A sailor presses down on the deck with a force of 600 newtons. How much work does he do in climbing 10 metres up the rigging?

3 A ship's anchor exerts a force of 5000 newtons. How much work will be done in lifting the anchor 30 metres?

How do Submarines DIVE and SURFACE?

SECTION EIGHT

Up and Down

1 With a bottle and some plastic tubing you can make a model 'submarine' that will surface at your command.

You can do this experiment in a bucket.

2 If you have the opportunity you could repeat the experiment using a long length of tubing at various depths in a swimming pool.

What differences do you notice as the depth increases? What blows the water out in a real submarine?

two-hole pung
glass tube
bottle
plastic tube
filled bottle
plastic tube

3 You could also experiment with working model submarines from toy shops.

4 In the enclosed space of a submarine there is the problem of there being enough air for the crew to breathe and keep alive.
How do submarines keep their fresh air?

How much air will a hundred-crew submarine need in one hour?
How much air do you take in during one breath?
Can you devise an experiment to measure your lung capacity?
By measuring your rate of breathing you could then work out how much air one person breathes (then how much 100 people breathe) in one hour.

You might like to take this opportunity to do some research into breathing and find out about our air:

1 Devise experiments to find how much of the air is used in breathing.

2 Devise experiments to find the proportion of air used in burning.

3 Does rusting use air – if so, in what proportion?

4 Your teacher may help you to make some oxygen and investigate its properties.

5 You could try some experiments making and using carbon dioxide.

6 How do green plants use air?

The Periscope

prisms

lenses

How does the captain see other ships when the submarine is below the surface?
You have probably heard of a periscope. How does it work?

A periscope uses mirrors (or prisms) and lenses (see the picture on the right).

1 Investigate with two mirrors: Can you see your friend round a corner?
Can you sit on the floor and see an object on the table?
Can you see over an obstacle?

2 It is fun to make and use a periscope. Before doing this it might be best to experiment and find the best position for the mirrors.

Here is one suggestion :—

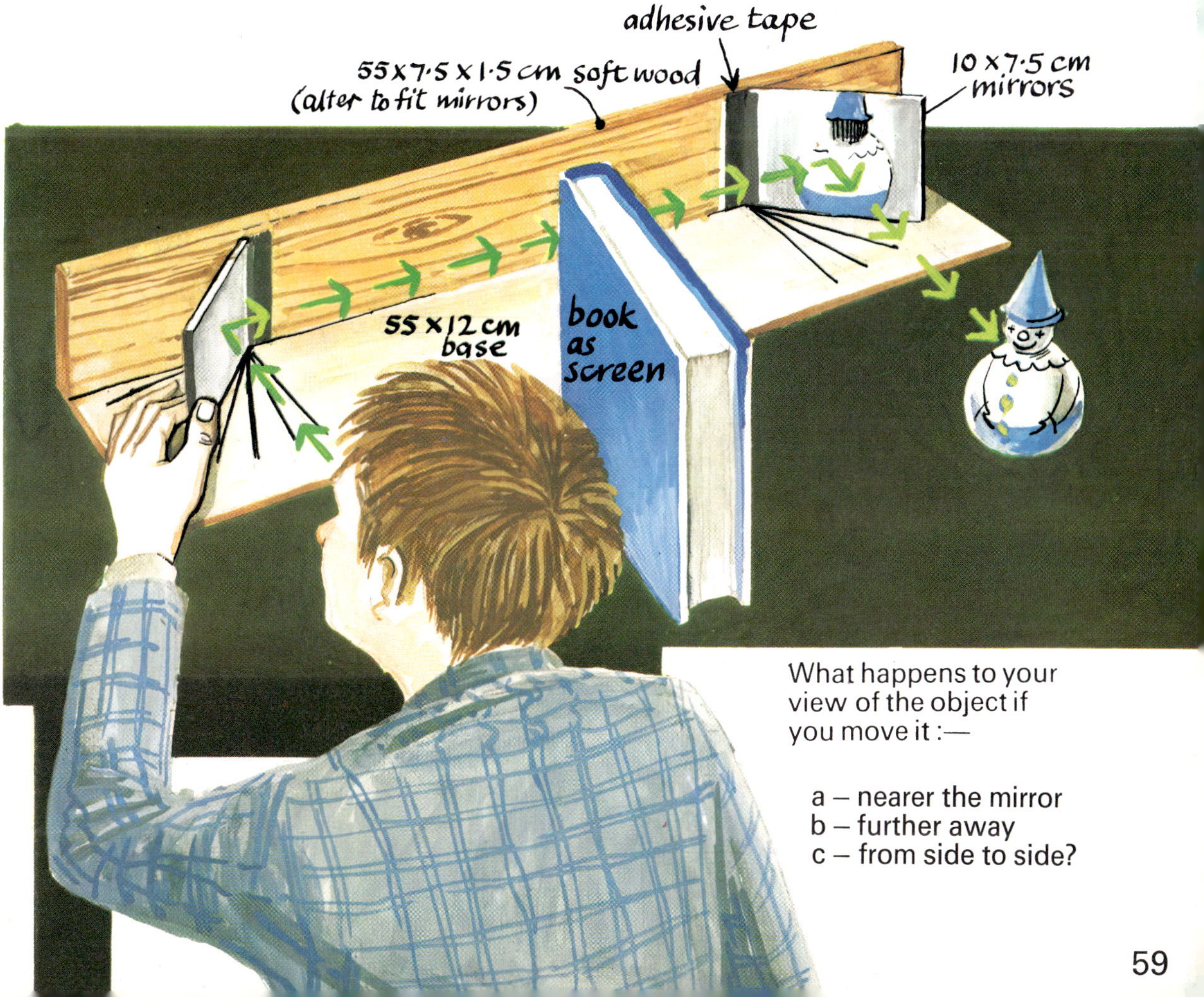

adhesive tape

55 × 7·5 × 1·5 cm soft wood
(alter to fit mirrors)

10 × 7·5 cm mirrors

55 × 12 cm base

book as screen

What happens to your view of the object if you move it :—

a – nearer the mirror
b – further away
c – from side to side?

Making a Periscope

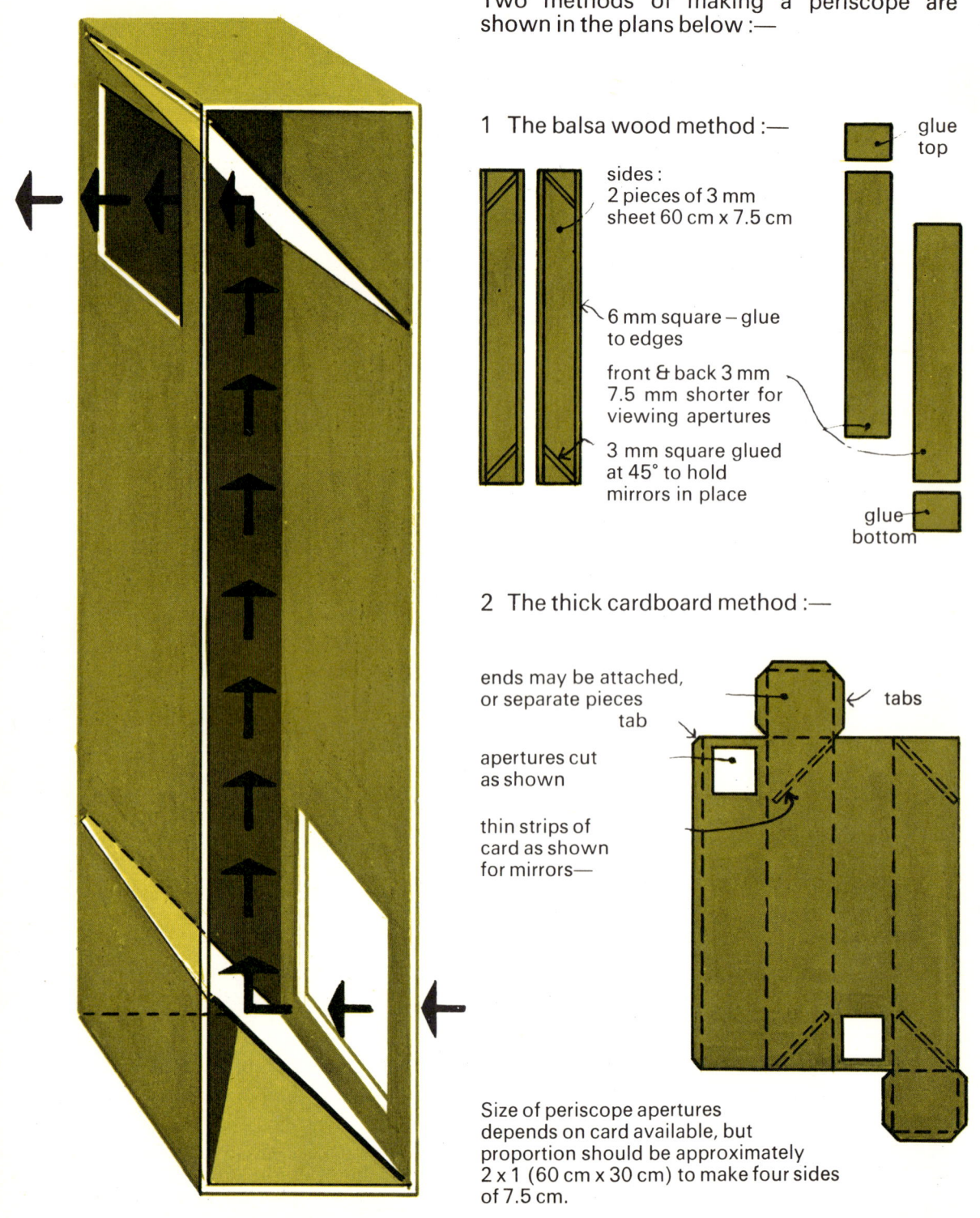

Two methods of making a periscope are shown in the plans below :—

1 The balsa wood method :—

glue top

sides :
2 pieces of 3 mm
sheet 60 cm x 7.5 cm

6 mm square – glue
to edges

front & back 3 mm
7.5 mm shorter for
viewing apertures

3 mm square glued
at 45° to hold
mirrors in place

glue
bottom

2 The thick cardboard method :—

ends may be attached,
or separate pieces
tab

tabs

apertures cut
as shown

thin strips of
card as shown
for mirrors—

Size of periscope apertures
depends on card available, but
proportion should be approximately
2 x 1 (60 cm x 30 cm) to make four sides
of 7.5 cm.

A cornflake pack will make a small periscope of 30 cms.

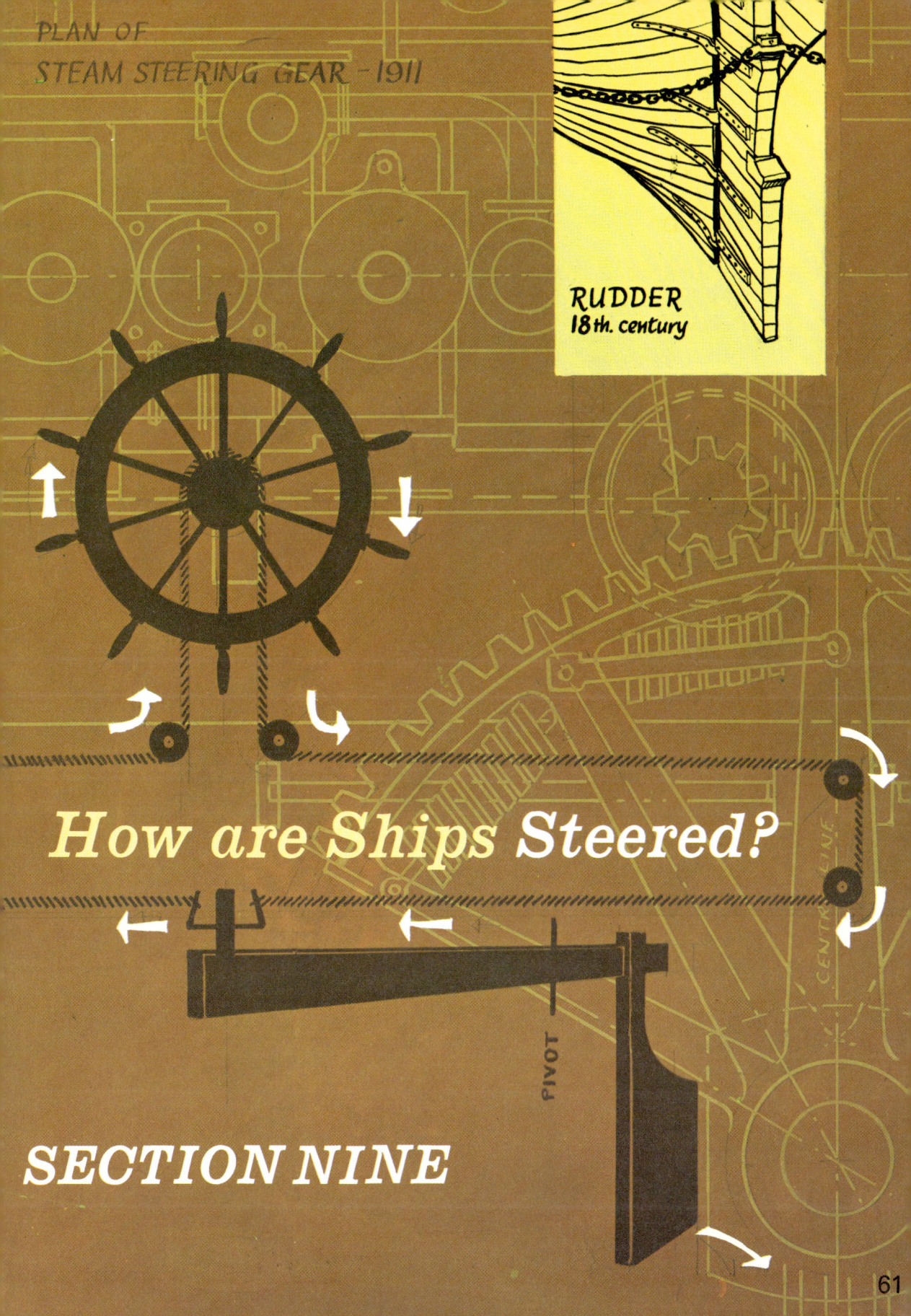

PLAN OF
STEAM STEERING GEAR - 1911

RUDDER
18th. century

How are Ships Steered?

SECTION NINE

PIVOT

CENTRE LINE

Steering your model

How is the ship controlled?

How is it kept on a straight course?
How does it turn to a new course?

Investigate the effect of a rudder on one of your model boats.

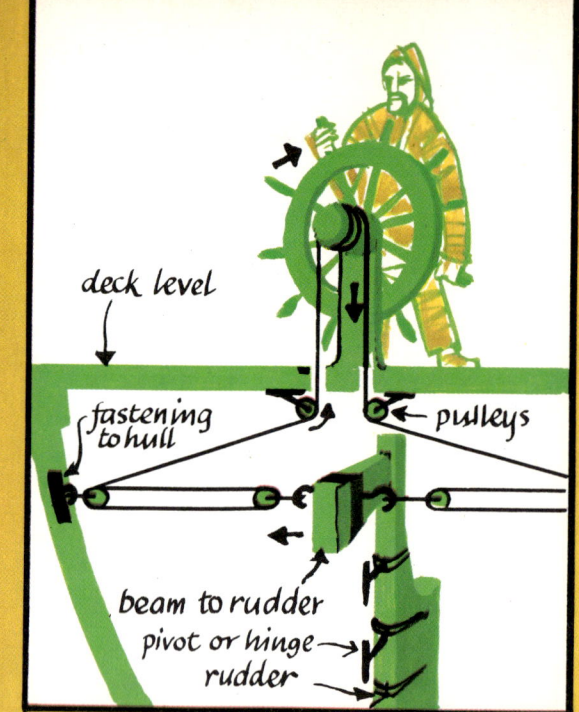

deck level

fastening to hull

pulleys

beam to rudder

pivot or hinge

rudder

You can use the models you have have made and add rudders:

Try bending the sheet tin rudder into different positions:—

How is the rudder moved?

Tiller

We can see how the tiller moves the rudder – but...

Steering wheel

... can you see how the rotary motion is transferred to the rudder? (Look at the picture above.)

How did Sir Francis Chichester steer Gypsy Moth IV during his lone voyage around the world? How does self-steering gear work? Find out about the terms 'starboard' and 'port', and how they relate to the placing of the tiller.

Two Special Ships

LIGHTSHIP

Can you devise a working electrical model of a lightship?

LIFEBOAT

Can you devise a boat that will always right itself if capsized and pushed under?

Questions and Answers

You must now realise what a vast topic for research the theme of 'Ships' is.

We hope that the questions in this book may have led you to ask others of your own, and that you have devised ways of finding some answers.

Finding answers to questions is most important.

Scientists rely on experiments to answer their questions.

Explore and ask why — then devise ways to seek out the truth.

This is what scientists do. The flow diagram on the next page may help you ask more questions.

An Integrated Study of Ships

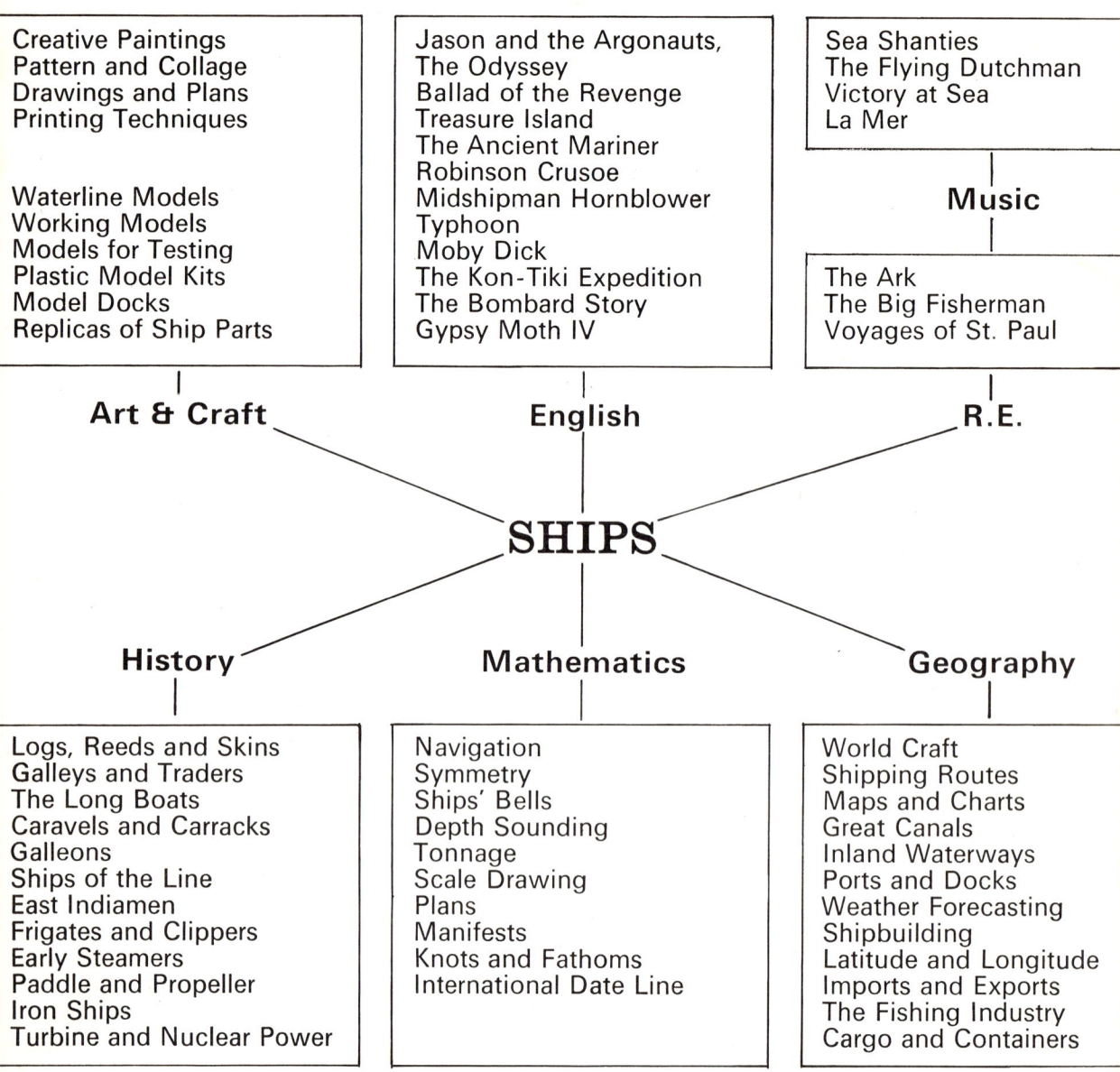

Art & Craft

Creative Paintings
Pattern and Collage
Drawings and Plans
Printing Techniques

Waterline Models
Working Models
Models for Testing
Plastic Model Kits
Model Docks
Replicas of Ship Parts

English

Jason and the Argonauts,
The Odyssey
Ballad of the Revenge
Treasure Island
The Ancient Mariner
Robinson Crusoe
Midshipman Hornblower
Typhoon
Moby Dick
The Kon-Tiki Expedition
The Bombard Story
Gypsy Moth IV

R.E.

Sea Shanties
The Flying Dutchman
Victory at Sea
La Mer

Music

The Ark
The Big Fisherman
Voyages of St. Paul

SHIPS

History

Logs, Reeds and Skins
Galleys and Traders
The Long Boats
Caravels and Carracks
Galleons
Ships of the Line
East Indiamen
Frigates and Clippers
Early Steamers
Paddle and Propeller
Iron Ships
Turbine and Nuclear Power

Mathematics

Navigation
Symmetry
Ships' Bells
Depth Sounding
Tonnage
Scale Drawing
Plans
Manifests
Knots and Fathoms
International Date Line

Geography

World Craft
Shipping Routes
Maps and Charts
Great Canals
Inland Waterways
Ports and Docks
Weather Forecasting
Shipbuilding
Latitude and Longitude
Imports and Exports
The Fishing Industry
Cargo and Containers

ACKNOWLEDGEMENTS

The authors are grateful to the following for permission to reproduce copyright photographs:
The Director, National Maritime Museum, Greenwich;
The Crown Copyright picture of the 'Ark Royal' by kind permission of The Controller of Her Majesty's Stationery Office;
T.H.I. Group Services Ltd.; Mr. K. J. Beken; Vickers Ltd.;
British Steel Corporation; Mr. John Etches; British Transport Films; Ministry of Tourism, Jerusalem, Israel.